CAMBRIDGE COUNTY GEOGRAPHIES

SCOTLAND

General Editor: W. Murison, M.A.

T0352055

PERTHSHIRE

Cambridge County Geographies

PERTHSHIRE

by

PETER MACNAIR, F.R.S.E., F.G.S.

Curator of the Natural History Collections in the Glasgow Museums
Lecturer on Mineralogy and Geology in the Technical College, Glasgow

With Maps, Diagrams and Illustrations

Cambridge:
at the University Press
1912

CAMBRIDGE UNIVERSITY PRESS
Cambridge, New York, Melbourne, Madrid, Cape Town,
Singapore, São Paulo, Delhi, Mexico City

Cambridge University Press
The Edinburgh Building, Cambridge CB2 8RU, UK

Published in the United States of America by Cambridge University Press, New York

www.cambridge.org
Information on this title: www.cambridge.org/9781107670792

First published 1912
First paperback edition 2013

A catalogue record for this publication is available from the British Library

ISBN 978-1-107-67079-2 Paperback

PREFACE

FOR their kindness in supplying various photographs reproduced in this volume I have to thank Mr James W. Reoch, Mr W. Lamond Howie, Mr George Herriot, Mr Charles Kirk, Mr John Annan, Mr James S. Boyd, and Mr P. D. Malloch. For the folding panorama of the Grampians, opposite p. 12, my thanks are due to Mr John Ritchie. For permission to photograph the bronze weapons on p. 102 I am indebted to Mr Ludovic MacLellan Mann. For many valuable suggestions and aid in connection with the book I have to acknowledge the assistance of my chief, Mr James Paton, of my colleagues Mr John Fleming and Mr David Gourlay, and of Mr James Park. My thanks are also due to Mr J. W. Reoch for the revision of the final proofs.

P. M.

January 1912.

CONTENTS

ILLUSTRATIONS

MAPS

The illustrations on pp. 4, 6, 8, 13, 20, 22, 24, 28, 30, 42, 47, 48, 80, 83, 98, 104, 107, 108, 112, 114, 118, 119, 120, 122, 123, 126, 127, 129, 130, 131, 133, 139, 149, 156, 157, 159, 160, 162,

*Map now available for download at www.cambridge.org/9781107670792

165, 170, 172, 173 are from photographs by Messrs J. Valentine & Sons; those on pp. 26, 115, 144, and 175 were supplied by Messrs F Frith & Co.; those on pp. 146, 152, 154 are from photographs by Messrs Annan & Sons; that on p. 97 was reproduced from Dr Hume Brown's *School History of Scotland* by permission of Messrs Oliver & Boyd; that on p. 77 is from the *Encyclopaedia Britannica* (11th Edition); that on p. 84 was kindly supplied by Messrs Pullar & Co.

NOTE. As this book was originally published before the full details of the 1911 census were obtainable, the following complete list of population figures is now given.

Berkshire 280,794

Abingdon 6809	Cumnor 1103
Aldermaston 559	Didcot 707
Aldworth 234	Donnington 626
Appleton 493	Earley 456
Ardington 471	Easthampstead 1959
Ashbury 660	Englefield 299
Avington 113	Faringdon 3079
Balking 190	Finchampstead 866
Basildon 596	Hagbourne, East and West 1430
Beech Hill 208	Hampstead Marshall 239
Beedon 238	Hampstead Norris 1217
Binfield 1912	Hendred, East and West 1066
Bisham 761	Hungerford 3040
Boxford 516	Hurley 1290
Bradfield 1606	Hurst 1069
Bray 3409	Ilsley, East 445
Buckland 682	Inkpen 693
Bucklebury 1136	Kintbury 1737
Burghfield 1343	Lambourn 2336
Chieveley 1066	Lockinge, East 264
Cholsey 2248	Maidenhead 15,219
Clewer 6356	Marcham 692
Coleshill 312	Newbury 12,107
Cookham 4915	Pangbourne 1677
Coxwell, Great 284	Radley 927
Crowthorne 3936	Reading 87,693

Sandhurst 3265
Shefford, Great, or West Shefford 398
Shinfield 2372
Shottesbrook 181
Shrivenham 602
Sonning 418
Sparsholt 335
Stanford in the Vale 859
Steventon 811
Stratfield Mortimer 1423
Streatley 732
Sunningdale 1537
Sunninghill 5335
Sunningwell 425
Sutton Courtney 790
Swallowfield 1533
Thatcham 2416

Tidmarsh 158
Tilehurst 480[1]
Twyford 1157
Uffington 523
Upton 209
Wallingford 2716
Waltham St Lawrence 937
Wantage 3628
Warfield 2283
Wargrave 2112
White Waltham 818
Windsor, New 12,681
Windsor, Old 2142
Winkfield 4193
Wittenham, Long 472
Wokingham 4352
Wytham 222
Yattendon 293

[1] The large decrease in the population of Tilehurst is·due to the transference of a large portion of it to Reading County Borough in 1911.

1. County and Shire. Origin and Meaning of Perthshire.

The term "shire" is derived from Anglo-Saxon *scir*, an administrative division presided over by the ealdorman and the sheriff (the shire-reeve). The term "county," on the other hand, arose after William I conquered England, when the lands were taken from the English earls and given to William's companions or *comites*. Each district was called a *comitatus* and from this we get the word "county." Like a great many other social institutions this division of our country into shires has been popularly attributed to the wisdom of some of our early rulers, King Alfred in particular being supposed to have taken an important part in the apportioning out of the country. It appears to be tolerably certain, however, that this theory of the origin of the different shires is exactly the reverse of what actually took place, the county not having been formed by the division of the country as a whole but by the aggregation of certain portions so as to form a county. From this point of view the county is simply the representative of a small community that has been merged into the unity of Great Britain. This opinion seems to be fully borne out by a consideration of many of our most important counties. It can also be shown that the county has been formed in a similar way by the aggregation of parishes. The parish, the manor and the township are traceable to independent tribal settlement. From this it will be seen that our counties have gradually grown up under varying conditions, and the boundaries have probably been shifted many times. In many cases the boundaries have been fixed by such a physical feature as the watershed of the country, this being easily recognised and utilised as a barrier between the adjacent divisions.

The origin of the name Perth is not very clear. Boece thought that it was derived from the Gaelic *Bar tatha*, "height of the Tay," referring to Kinnoull Hill, which rises abruptly from the Tay to the east of the city. On the other hand Stokes, who is probably right, makes it Pictish *perth*, "a thicket," and neither height over the Tay, nor confluence of the Tay, *Aber tatha*, as maintained by those who consider that the town was originally situated at the confluence of the Almond with the Tay.

It ought to be stated at the very outset that the great factor which has determined the present geographical conditions of Perthshire has been the Highland boundary fault or line of demarcation between the highland and lowland portions of the county. In the course of these pages we hope to be able to show that not only are the scenic and physiographical features of the shire directly due to the different geological structure of these two great natural divisions, but also that its soils, climate, natural history, agriculture, population, the distribution of its towns and villages, its people, their language and their history, have largely been determined by this all important factor.

The shire lies in one compact mass. Formerly it had two small detached portions in the south, on the Forth. One of these was included in the parish of Kippen, which lay wholly across the Forth, while the other embraced the parishes of Culross and Tulliallan now in Fifeshire.

In the neighbourhood of Stirling Logie parish enclosed a detached portion of Fifeshire, and Collace parish near Perth a portion of Forfarshire. Many of these anomalies have recently been done away with. How they originally came to be arranged in this whimsical fashion is not easy of explanation; but it is supposed that when the counties were being formed the landlords put their lands into those districts in which they had the greatest interest. The origin of the parish boundaries is equally difficult of explanation as many of them are very irregular and appear to be of a purely arbitrary character.

2. General Characteristics and Natural Conditions.

The county of Perth is situated in the middle of Scotland and, with the exception of the small tidal tract represented by the alluvial flat that lies between the Sidlaws and the sea, known as the Carse of Gowrie, is wholly an inland county.

Perthshire is bounded on the north-west by Inverness-shire, on the north by Inverness-shire and Aberdeenshire,

Firth of Tay

on the east by Forfarshire, on the south-east by Fifeshire and Kinross-shire, on the south by Clackmannanshire and Stirlingshire, on the south-west by Stirlingshire and Dumbartonshire, and on the west by Argyllshire.

Sir Walter Scott in the *Fair Maid of Perth* says, "Amid all the provinces in Scotland if an intelligent stranger were asked to describe the most varied and the most beautiful it is probable he would name the county of Perth. A native also of any other district of Caledonia though his partialities might lead him to prefer his native county in the first instance would certainly class that of Perth in the second, and thus give its inhabitants a fair right to plead that—prejudice apart—Perthshire forms the fairest portion of the northern kingdom."

Perthshire affords examples of the most romantic and grandest scenery in Scotland, much of which has been rendered classic by important events in Scottish history. Mountains, lakes, rivers, cascades, woods and rocks supply the elements that combine to make up all that is grand and beautiful in every landscape. In the course of a few miles one may pass from a deep ravine or rugged Alpine glen into a rich and open valley which partakes of the cultivated beauty of the lowlands and in the centre of which lie embosomed the waters of a great lake. Or one may follow the wanderings of a great river from its source among the mountains, whence, as a torrent and with a wild mountain cry, it precipitates itself over ledges of rock to become lost on the black moor beneath but after a course of many miles finds itself meandering through a spacious vale or widespread wooded plain.

Falls on the Dochart

Geographically the mainland of Scotland can be divided into three parts, the Highlands, the Southern Uplands, and the Midland Valley, each characterised by a particular set of rocks and by a scenic aspect which is intimately connected with its geological structure. The dividing line between the Highlands and the Midland Valley, known as the great Highland boundary fault, crosses Scotland from shore to shore with a north-east and south-west trend. Geographically it divides the Highlands from the Lowlands and geologically the crystalline schists from the Old Red Sandstone. The position of this great line of demarcation has been more or less accurately fixed. It can be traced through Arran and Bute, thence from near Toward Castle to Innellan and across the eastern point of Rosneath Peninsula, and by Helensburgh across Loch Lomond to Balmaha. It enters Perthshire at Aberfoyle, passing through Callander, Comrie, Crieff, Birnam, Blairgowrie to the Bridge of Cally and Alyth, where it leaves the county, striking north-eastwards to the sea at Stonehaven.

Situated as it is upon this great divisional line, Perthshire is divided into two distinct regions—the Highlands and the Lowlands. The greater part of the Highland region is open moorland; large tracts of it, however, have been planted with larch and Scots fir. The Lowland region on the other hand is noted for its fertility, notably the valley of Strathmore and the Carse of Gowrie. The greater part of the county, however, is wholly unfit for the raising of grain or green crops, only about one-fifth of the entire area being cultivated.

With only a few exceptions the rivers and streams flow in a south-easterly direction, and reach the ocean by the way of the Firth of Tay or the Firth of Forth. As a rule they issue from large elongated lochs situated in the main valleys.

Its position in the very heart of Scotland has made Perthshire the scene of some of the most important and

Comrie

stirring events in Scottish history, and almost every part of the shire is connected in some way with the past history of the country.

The great divisional line just referred to was that which originally separated the Celtic natives from the invading hordes from across the North Sea; and to this day it serves to mark off the areas occupied by the Gaelic-speaking and the English-speaking people. In the Lowland region we hear only English spoken, often with a strong northern accent. Scattered here and there over the great plain of Strathmore are numerous villages and towns, the houses of which are usually well built of solid stone and lime, and roofed with flagstones, slates, or thatch. The common fuel is coal brought by land or sea from the south. Immediately we pass to the north of the great boundary line, we meet with a totally different condition of things. The Gaelic language is now the characteristic tongue. Villages are few, and the houses are built simply of unhewn boulders taken from the surrounding fields, the binding materials being merely clay or earth. The interiors are of the simplest character and peat is largely used as fuel. That these features have been modified to some extent by the recent development of railways in the Highlands must be admitted, but the general contrast is still quite sufficient to mark off the one region from the other.

In the Highlands the principal villages are situated either at the ends of the lochs or at some favourable point in the main valleys, while along the margin of the Highlands the villages have usually been built where the valleys open to the plain, as at Crieff and Callander.

It is worthy of note that at the beginning of the nineteenth century Perthshire was the second most populous county in Scotland, Lanark having then only 22,000 more inhabitants than Perth, whereas now Lanark has 1,216,000 more inhabitants than Perth, and Perth stands ninth in point of population. The reason why Perth has made no progress is not far to seek, and is simply due to the fact that Perthshire is entirely outside the bounds of the Carboniferous Formation, whose mineral wealth has been the great factor in the rapid rise and development of other counties during the last hundred years.

3. Size. Shape. Boundaries.

The county of Perth lies between 56° 7′ and 56° 57′ N. latitude and between 3° 4′ and 4° 50′ west longitude. In size it is the fourth largest county in Scotland. From east to west, its greatest length is about 70 miles, and its greatest breadth from north to south about 56 miles. Its total area is something like 2500 square miles.

At some parts the boundaries are natural and well defined, while at others they are purely artificial and not so easy of definition. Beginning near Perth, the boundary line can be traced along the north bank of the Tay as far as Invergowrie, where it bends sharply northward and then westward. It then follows a somewhat arbitrary course, successively passing through or near Coupar-Angus, Alyth and Airlie: thence it proceeds along the western watershed of Glen Isla. From that point it crosses a number of summits and saddle points including the Cairnwell Pass, over which the road to Braemar passes. This is probably the highest driving road regularly used in Great Britain. The boundary line can now be traced westward by the head of Glen Tilt, where it meets the junction of Aberdeen and Inverness at an altitude of 3267 feet and overlooks the headwaters of the infant Dee. Continuing

in a westerly direction, it never drops lower than 3000 feet
until it reaches Lochan Duin to the west of Glen Bruar.
Still further west it crosses the Highland Railway line a
little to the north of Dalnaspidal, at an altitude of 1454
feet above the level of the sea.   The boundary now follows
the summits which lie to the east of Loch Ericht, continuing to fall till it reaches the level of that loch at an
altitude of 1153 feet.   Skirting the south side of Ben
Alder, it passes across the Moor of Rannoch, and, keeping to the highest ground, intersects the West Highland
Railway near the headwaters of the Leven, and shortly
after marches with the county of Argyll.   From this
point it cuts successively across the summits of Ben
Creachan, Ben Achallader, Ben-a-Chaisteil and Ben
Odhar, till it reaches the watershed between the river
Lochy and the river Fillan.   Thence it mounts the
summit of Ben Laoigh (Ben Lui), having on the east the
infant Tay, here known as the Coninish Water.   This is
the extreme western point of the boundary line, which
now turns east marching with Dumbartonshire.   It crosses
the Falloch at Inverarnan and the West Highland Railway a little further on.   Mounting again to the summits,
it crosses to Glen Gyle, where it joins Stirlingshire.   From
thence it passes the head of Loch Katrine and skirts the
east side of Loch Arklet, which is now being actively
prepared as an addition to the Glasgow water supply.
Keeping to the east side of Ben Lomond, it descends
the Duchray Water to the neighbourhood of Aberfoyle.
Striking eastward, it follows the line of the Forth to its
junction with the Allan Water about two miles from

Stirling; and just excluding Bridge of Allan, it sweeps past Sheriffmuir to Clackmannanshire. The boundary now crosses the Ochils to the neighbourhood of Dollar. Proceeding by the Yetts of Muckart and Fossaway, and keeping the high ground between Dunning and Milna-thort, it touches Kinross. From this point it strikes in a north-easterly direction across Glen Farg to the west of Newburgh on the Firth of Tay. It then bends sharply west along the south bank of the Tay to the Bridge of Earn, the point from which we started.

Roughly then the boundary line of Perthshire may be defined as an irregular circle with its centre near the head of Glenalmond, and having a radius of about 32 miles and a circumference of over 300 miles.

4. Surface and General Features.

Perthshire is wholly an inland county with the exception of the small maritime tract between Perth and Invergowrie, known as the Carse of Gowrie. The county can be divided into two distinct parts, namely, the Highland region, which forms the north-western portion, and the Lowland region, which forms the south-eastern portion. The Grampian mountains, which cor-respond to the Highland portion of the county, enter it at the north-east corner. At that point they simply form the northern boundary line of the shire. But as they are traced westwards they spread further and further into the county till they practically occupy the whole of it from north to south.   On the other hand, the Lowland division is broadest in the north-east and, when traced westwards, passes almost entirely out of the county.   The southern boundary of the county between Invergowrie on the Firth of Tay to Stirling is marked by the Sidlaw and Ochil Hills.   In the Highland region the mountains rise to an average elevation of about

Ben More

3000 feet, while many of the peaks exceed this altitude. In the south-west corner of Perthshire the chief mountains are Ben Laoigh (3708 feet), Ben Odhar (2948), Ben More (3843).   On the ridge that separates Loch Tay and Glen Lyon are Ben Lawers (3984 feet), the highest mountain in Perthshire, Meall Garbh (3661), Meall nan Tarmachan (3421).   Further east and in the same line of bearing, Farragon Hill (2559) and Ben Vrackie (2757). In the north-east of the county, and along the boundary of Atholl are Carn an Fhidhleir (3276), An Sgarsoch (3300), Cairnwell (3059), Ben-y-Gloe (3671) and Ben Vuroch (2961). The principal mountains on the ridge that separates the river Lyon and Loch Rannoch, are Schiehallion (3547 feet), Carn Gorm (3370), Carn Mairg (3419); north of Loch Lydoch and Loch Rannoch, Ben Alder (3757), Carn Dearg (3084 feet); and in the north-west of Perthshire and the neighbourhood of Loch Lyon, Ben Creachan (3540 feet), Ben Heasgarnich (3530), Meall Ghaordie (3407), Ben Vannoch (3125). The belt of high ground forming the Sidlaw and the Ochil Hills is separated from the Grampians by the lordly valley of Strathmore. The Ochils lie in the south of Perthshire and stretch from the Forth near Stirling to the neighbourhood of Perth. Some of the principal heights are as follows: Mickle Corum (1955), Blairdenon Hill (2072), Core Hill (1780), East Bow Hill (1562), Carlowrie Hill (1552), Muckle Law (1306), Rossie Law (1064), Skymore Hill (1302), Cock Law (1337), and Castle Law (1028). The Sidlaw Hills on the north side of the Firth of Tay separate Strathmore from the Carse of Gowrie, and may be considered as simply a northern prolongation or branch of the Ochils. The principal altitudes taken in order from west to east are Kinnoull Hill (729), Evelick or Pole Hill (944), Black Hill (1182), Dunsinane Hill (1012), King's Seat (1235), Blacklaw Hill (929), and Balo Hill (1029).

The line of demarcation between the Highland and

Killin Hills

the Lowland region passes diagonally across the county in a north-east and south-west direction. It can be traced from near Alyth by the Bridge of Cally, Birnam, Bankfoot, Logiealmond, Comrie, Callander onwards to Aberfoyle. The region between this line and the Ochil and Sidlaw Hills forms the great valley of Strathmore. Orographically, then, Perthshire may be considered as consisting of three parallel bands or belts. The most northerly of these is a highly mountainous region and may be called the Grampian belt. To the south of this comes a broad plain or valley, the Strathmore belt. Still further south is the hilly ground which may be called the Ochil and Sidlaw belt. It will presently be shown that the rocks forming the valley of Strathmore and the Sidlaw and Ochil Hills belong to the same period in geological time, and though they vary somewhat in elevation are classed together as the Lowland part of the shire. On the other hand, the rocks lying to the north of the great line of demarcation just described are of a totally different character, belonging to a much older period in geological time, and forming the Highland area.

If the reader looks at the view taken from the summit of Ben Lawers it will at once be seen that the Grampians appear to form a great level plateau, deeply indented with valley systems. The use of the term plateau to describe what is generally looked upon as a mountainous country requires some explanation. This illustration shows the remarkable uniformity of level to which all the mountains rise, so that if we could imagine

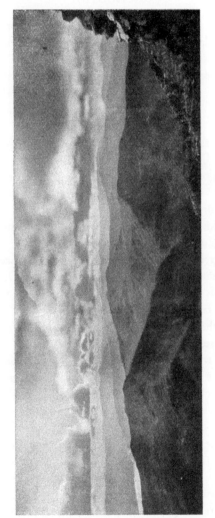

View from Summit of Ben Lawers

all the intervening valley systems filled up, there would
be a great plain rising to a uniform level of about 3000
feet above the sea. The origin of the plain will be
discussed presently. In the meantime its existence is
simply indicated that the reader may grasp the true
character of the mountainous region of Highland Perthshire.

If we stand on one of the eminences of the Ochil
or Sidlaw Hills and look across the valley of Strathmore,
we shall at once be struck with the long rampart of the
Grampians which rises abruptly from the plain and
forms the dividing line between the two great divisions
of Perthshire. This feature is strikingly shown in
illustration on p. 8.

The great valley of Strathmore lying between the
mountain-rampart of the Highlands and the Ochil and
Sidlaw Hills extends from Dumbartonshire on the west
to the German Ocean at Stonehaven on the east. It
enters Perthshire at the Bridge of Craithie near Meigle
and increases in breadth, scenery and variety of features,
to a point where the Isla joins the Tay near Kinclaven
Castle. At this point it reaches its noblest and most
impressive character, having a breadth ranging from 12
to 14 miles. To form anything like an adequate
conception of the greatness of this queen of Scottish
valleys, one must have climbed the Sidlaws or the
Grampians and looked down upon the far stretching
band of low country, remarkably beautiful and fertile,
and dotted with numerous towns, villages and mansions.
The picture on p. 160 shows the Strath from Birnam

Hill with the river Tay, which passing through the gateways of the hills has just escaped from its mountain barriers.

The Sidlaw and the Ochil Hills, bounding the southern part of the county, present a low chain of long round-backed swelling hills, covered with vegetation and in some cases under cultivation up to their summits. Numerous defiles or passes intersect the chain, cutting it into smaller masses and single hills.

If you look at the orographical map at the beginning of the volume you will at once see that the valley systems of Highland Perthshire fall naturally into two great classes, namely, the longitudinal and the transverse. The longitudinal valleys have a general north-east and south-west trend and coincide roughly with the strike and outcrop of the rocks of the Highland area. The following are examples of the valleys of this type—Glen Dochart, the valley of Loch Tay, the Tay between Kenmore and Ballinluig, Glen Lyon, the valley of the Tummel, Loch Earn, and the valley of Loch Katrine. On the other hand, the transverse valleys cut across the strike and outcrops of the rocks, crossing the longitudinal valleys approximately at right angles. The valleys of the Shee, the Isla, the Ardle, and the Garry are examples of the transverse type.

The Carse of Gowrie is a low tract of alluvial land and stretches from Kinnoull Hill to Invergowrie, having a total length of about 15 miles and varying in breadth from two to four miles, with an altitude of from 24 to 40 feet above sea-level. Practically the whole of the

Dollar and the Ochil Hills

Carse consists of rich arable land covered in the summer time with broad fields of corn and extensive orchards and dotted here and there with houses, proprietorial mansions and a few villages.   The Carse of Gowrie has been fitly called the Garden of Scotland.

The deer-forests of Perthshire, six in number, are— Atholl containing 35,540 acres, Fealar 14,500 acres, Glen Bruar 11,000 acres, Drummond Hill 2400 acres, Glenartney 19,310 acres, and Rannoch 12,000 acres. From this it will be seen that a very large proportion of the county is covered with deer-forests.   They contain large numbers of red deer, roe deer, and fallow deer.   According to the *Sportsman's and Tourist's Guide* for 1911 the rental of deer-forests in Perthshire is £75,000. The grouse-moors in the county, of which there are over 400, are unsurpassed and yield magnificent sport.

5. Watershed. Rivers. Lakes.

It has already been pointed out that on the west and the north the watershed coincides pretty closely with the boundary line of the county, generally dividing the headwaters of the Stirlingshire, Dumbartonshire, Argyllshire, Inverness-shire and Aberdeenshire streams from those of Perthshire.

The Tay, which is the longest river in Scotland, rises in a corrie on the north side of Ben Laoigh on the confines of Argyllshire and Perthshire at an altitude of 3000 feet above sea-level.   From its source to Loch

Dochart is a distance of 11 miles and in this part of its course, where it is called the Fillan Water, it has fallen 500 feet. Then it passes through Loch Dochart and Loch Iubhair. For 14½ miles it is known as the Dochart, and ultimately falls into Loch Tay at Killin. Including the Fillan and Dochart, the river Tay has a total length of about 117 miles, and drains an area of

Near the Source of the Tay

close on 2000 square miles. The chief sections of the river may be summarised as follows :

Source to Loch Tay	25	miles
Head of Loch to Kenmore	14½	,,
Kenmore to junction with Tummel	15½	,,
Junction of Tummel to Perth Bridge	31	,,
Perth to mouth	31	,,

The gradient of the Tay between Loch Dochart and Loch Tay is comparatively slight, Loch Tay being 350 feet above the level of the sea. At the confluence of the Tummel and the Tay it has fallen to an elevation of 200 feet, and near Perth the elevation may be said to have disappeared as the river has now become tidal.

The valley of the Tay from Dunkeld to Kenmore for a space of 25 miles is a continued scene of unsurpassed beauty and loveliness. Here the majestic river winds through a richly wooded and cultivated region, bounded on each side with lofty mountains. It is joined on its left bank a few miles below Kenmore by the Lyon, which rises in Loch Lyon ; and near Ballinluig Station, by the Tummel from the Moor of Rannoch. The Tummel drains Loch Lydoch, Loch Ericht, Loch Rannoch and Loch Tummel, and also brings with it the waters of the Garry from Loch Garry. Near Dunkeld, the river receives on its right bank the Bran, draining Loch Freuchie ; opposite Kinclaven Castle on its left bank, the Isla enters, bringing with it the Shee, the Ardle and the Ericht. Two miles above Perth, the river Almond, which rises to the south of Loch Tay, joins the main stream opposite Scone Palace, while below Perth comes in the Earn on the right bank, the last tributary of any importance. If the river has lost the picturesqueness of its highland course in the noble curve with which it sweeps across the valley of Strathmore, this is more than balanced by the gain in majesty from the many tributaries just described.   The

Glen Dochart

sudden changes which the river makes in its course from its source to the sea are full of great interest.

The longitudinal valleys of the Earn, the Almond, the Bran, the Tay and the Tummel are terminated at their eastern extremities by a great transverse valley which, running in a north-west and south-east direction, and descending from the very heart of the mountains, has cut off the longitudinal valleys nearly at right angles. This valley is occupied by the Garry, which, as it sweeps onwards, gathers up the waters of the various longitudinal valleys, carrying them southwards in a combined stream. This great transverse valley terminates at Dunkeld, where the river emerges from the Highlands on to the valley of Strathmore. Now liberated from the narrow mountain barriers by which it was hitherto confined, it assumes a more winding course but the general trend is still towards the south as far as Perth. At this point the river meets with a formidable barrier in the Sidlaw Hills. This, however, it has been able to breach between the hills of Moncrieff and Kinnoull. It was from a point on the former of these eminences that the Romans were supposed to have caught their first glimpse of the Tay, when they exclaimed in rapture *Ecce Tiber ! Ecce Campus Martius !* "Behold the Tiber ! Behold the field of Mars !" The exclamation was more complimentary to the Tiber than the Tay; or, as Sir Walter Scott puts it,

> "'Behold the Tiber!' the vain Roman cried,
> Viewing the ample Tay from Baiglie's side ;
> But where's the Scot that would the vaunt repay,
> And hail the puny Tiber for the Tay?"

Kinnoull Hill and the Valley of the Tay

The river Forth belongs to the south-east corner of the shire and shows how the Highland region has crept towards the south. The Forth can be considered a Perthshire river only in the same sense as a man born in Perthshire but spending the greater part of his life outside the county can be spoken of as a Perthshire man. The Avondhu and the Duchray, the headwaters of the Forth, rise on the east side of Ben Lomond at an altitude of over 2000 feet. These two streams run in a parallel direction to the south-east, the Duchray water forming the boundary between Perthshire and Stirlingshire, and the Avondhu flowing through Loch Chon and Loch Ard. The streams meet a little to the west of Aberfoyle and just before they pass on to that portion of Strathmore formed by the valley of the infant Forth. The river now meanders, coquetting between the shires of Perth and Stirling but finally abandoning the county of its birth. East of this the Forth receives the following tributaries on the left or Perthshire bank—the Goodie, the Teith, the Allan and the Devon. The Teith, like the Forth, rises in two headwaters, one of which descends from the south side of Ben a Chroin and flows through Loch Voil and Loch Lubnaig, being successively known as the Balvaig and the Leny. The other flows from Loch Katrine through Loch Achray and Loch Vennachar and unites with the Leny at Callander.

In many respects that part of Perthshire drained by the basin of the Forth is the most interesting and picturesque part of the county. The southern stream after emerging from Loch Katrine begins to traverse the

Trossachs, round which the great Wizard of the North has thrown such a halo of romance. It is flanked on the north by Ben A'an and on the south-west by Ben Venue, great mountain masses which rise tier upon tier in a series of rocky eminences of the most fantastic character from the pass below. The whole of the lower ground is covered with a dense growth of herbs, shrubs

Loch Katrine

and such trees as hazels, oaks, birches, hawthorns and mountain ashes.

In close association with the rivers are the lakes of Perthshire, which are numerous, large and renowned for their natural beauty. For the most part these lakes are confined to the northern or Highland division of the county. They often appear in linear groups like so many diamonds strung upon a thread of silver. The largest lochs in Perthshire are Loch Tay, Loch Earn and Loch Rannoch in Breadalbane, Loch Ericht on the confines of Perthshire and Inverness-shire, and Loch Katrine in the district of Menteith. These are followed in size by Lochs Lydoch on the confines between Perth and Argyll, Garry between Rannoch and Atholl, and Tummel in Atholl. In the south-western part of the county we have Loch Lubnaig, Loch Voil, Loch Vennachar and the Lake of Menteith. Innumerable smaller lakes, principally confined to the Highland region, need not be mentioned in detail.

Classified according to their origin and mode of occurrence, the lakes of Perthshire can be arranged into three distinct types. The first have been hollowed out of the solid rock and are known as true rock basins. These include all the larger and more important lochs, such as Loch Tay, Loch Earn and Loch Rannoch. Second are those which have been formed by the ponding back of a sheet of water by glacier *débris*. This type is usually confined to the heads of glens or the mouths of corries. These lochs are not usually of any great size but they occur in great numbers in the Highlands. The third type includes all those which lie in cup-like hollows either in the old glacier moraines or in the boulder clay. Fine examples occur in the area between Dunkeld and Blairgowrie, where there is a chain of them including the Loch of Lows, Butterstone Loch, Clunie Loch, Marlie Loch and Rae Loch.

The origin of the second and the third class of lake

is so evident as to require no explanation, but those of
the first order are more difficult to explain. The view
now most generally accepted is that first advanced by
Sir A. C. Ramsay, who accounted for such great rock-
basins as Loch Tay and Loch Earn by the theory that
they had been scraped out by the agency of ice. Everywhere the sides of these rock basins show beautifully

Loch Tay

smoothed and scored surfaces as if some tremendous
weight had passed over them, grinding and polishing
them in its onward march. Now the only agent that
we know to have been in operation during past ages
which would be sufficient to account for such phenomena
is ice. It was partly during the vast extension of the
ice sheet and partly during the later valley glaciation,

that the rock basins which enclose our Highland lochs
were excavated.

Within the catchment area of the Tay there are many
different types of rock basins, the simplest of these being
that of Loch Earn. This loch has a length of over six
miles and an average width of about three-quarters of a
mile. The maximum depth, 287 feet, occurs half way
down the loch. A great fault enters the loch at Glen
Ample and crosses it diagonally to Dalveich. This fault
coincides with a small basin which has a depth of 240 feet.
It has been shown that during the period of maximum
glaciation the ice-sheet crossed this part of Perthshire in an
east-south-east direction, and as a consequence the greatest
pressure must have been exerted on the south side of the
valley. This is confirmed by the fact that the slope
of the southern side of the valley is twice as steep as that
of the northern.

In the great rock basin of Loch Tay, which is 14½
miles in length, with an average breadth of three-quarters
of a mile and a depth of 510 feet, we have a somewhat
different type of basin from that of Loch Earn. In this
basin there appears to have been a deflection of the ice
towards the north-east in the neighbourhood of Ardeonaig
and Ardtalnaig and this accounts for the deepest part of
the basin being situated to the east of the latter village.

Two great sculptors, then, have been at work in
producing the river and lake system of Perthshire. The
first was the ordinary agent of sub-aerial denudation, the
second that of a great sheet of ice which has now entirely
disappeared from these islands. The work of the first

sculptor was to trench the old plateau of marine denudation into the great valley systems we have just described,
and derive from the monotonous table-land the picturesque
valleys and gorges that now diversify this part of Perthshire. The work of the second sculptor was to add the
charm of lake to that of stream and so complete the scenic
beauty of the Highlands of Perthshire.

6. Geology and Soil.

By examining the crust of the earth, geologists have
been enabled to classify the great rock masses of which it
is composed into two kinds, according to their mode of
origin—those which have been erupted from the interior
of the earth in a molten condition, known as igneous
rocks ; and those which have been formed as sediment at
the bottom of seas and lakes, and which have been piled
up into thick beds of strata, known as sedimentary rocks.
A third group, the metamorphic, is generally adopted by
geologists for convenience in the matter of description;
but, as it includes rocks, some of which were of igneous
and some of sedimentary origin, this classification is somewhat objectionable. In accordance, however, with general
usage, it has been followed in this description of the rocks
of Perthshire.

According to their mode of occurrence, the igneous
rocks may be subdivided into two groups. The first
comprises those which have been ejected upon the surface
of the earth, by volcanic action, and have been laid down

either as great sheets of lava, or accumulations of fragments of lava and volcanic dust. These are known as volcanic rocks. The Sidlaw and the Ochil Hills are formed of rocks of this kind. The others, called plutonic rocks, have cooled at some distance below the surface and have solidified much more slowly than volcanic rocks. As a result of this they have assumed a more coarsely crystalline structure. They commonly occur in great intrusive bosses. The granites and diorites of Glen Lednock, the Moor of Rannoch, and Glen Tilt may be taken as examples of this division of the igneous rocks.

The great series of sedimentary or fragmental rocks include all those which, like sandstone, have had a secondary or derivative origin, or, in other words, which have been formed out of previously existing materials, as well as a few others which, strictly speaking, do not answer to this description of their origin. Some of these have been formed by the action of wind along the sea coast, such as sand dunes. Others owe their origin to moving water, and under this category come gravel, sand and mud.

Another great division of the sedimentary rocks is that known as the organically formed rocks, which have been built by the slow accumulation of the remains of plants and animals existing upon the surface of the earth and in lakes or seas. Coal and limestone are familiar examples of this class. The great valley of Strathmore is paved with a vast thickness of sedimentary rocks, principally sandstones, shales and conglomerates of Lower Old Red Sandstone age.

It is often found that both igneous and sedimentary

rocks have been altered by pressure or by coming into contact with molten igneous material. In this way clay or shale may be altered into slate, and sandstone into quartzite, while a shaly sandstone may pass into mica schist. Such igneous rocks as granite become gneiss, and whinstone is altered into hornblende schist. When rocks have been subjected to such alterations they are known as metamorphic rocks.

The Highlands of Perthshire present a region in which all the phenomena connected with metamorphism may be studied in their most minute details. In this region there is a great series of sedimentary rocks which have been altered by metamorphism from such normal sediments as conglomerates, sandstones, shales and limestones into schistose conglomerates, quartzites, slates and crystalline limestones. These sediments prior to their metamorphism were penetrated by intrusive igneous rocks, which have also suffered in the general metamorphism, passing into gneisses and hornblende schists. At a later period the metamorphic rocks were invaded by great masses of igneous material, principally granites, which produced a still further stage of metamorphism along the line of contact.

All the sedimentary rocks show evidence of having been originally laid down in more or less horizontal beds or strata. They are no longer seen, however, to occupy the original horizontal position in which they were formed but have usually been bent into a series of folds as a result of the secular cooling of the earth's crust. When the strata form a series of undulations the hollows are called *synclines* and the ridges *anticlines*. The rocks forming the valley of Strathmore are arranged in a syncline, while those of the Ochil and the Sidlaw Hills have an anticlinal arrangement. In the Highlands the rocks have been so intensely folded that the synclines and anticlines have become closely packed together in such a way that the axes of the folds are no longer perpendicular but are seen to be inclined in a definite direction over great areas. This is known as the isoclinal type of folding.

The great line of demarcation to which we have already referred runs across Scotland in a diagonal direction, dividing the country into two portions—the Highlands and the Lowlands. This line is a geological as well as a geographical line, and separates the crystalline schists of the Highlands from the younger Palaeozoic rocks of the Midland valley. It is a line of fault, the rocks of the Midland valley having been thrown down for many thousands of feet against those of the Highlands.

The rocks which form the Highlands of Perthshire are metamorphic. In the majority of instances they were laid down as sedimentary deposits, subsequently altered, both by the great plication and pressure to which they were subjected, and by the intrusion of great bosses of igneous material. The stage of alteration exhibited by these metamorphic rocks varies to a considerable extent. In some of the more siliceous members the original grains of quartz are still easily recognisable, while in others the rock has become so reconstructed by metamorphism that the original character is no longer discernible.

The different schists which form the Highlands of Perthshire traverse the county in bands or zones, having a

3—2

general north-east and south-west trend, and may be said
to lie roughly parallel with the great boundary fault. The
following table shows the general succession of the zones
as they are traced from south to north.

	13. Moine Schists. (*a* on map.)	
	12. Quartzite and Quartz-schist, with pebbly conglomerate.	(*x* on map.)
	11. Schiehallion conglomerate ("Boulder Bed").	
	10. Limestone ("Blair Atholl"). (Blue on map.)	
CRYSTALLINE	9. Black Schist. (*g*^{1} on map.)	
SCHISTS	8. Phyllites etc. ("Ben Lawers Schist"), (*l*^{1} on map.)	
OF THE	7. Garnetiferous mica-schists. (*g* on map.)	
PERTHSHIRE	6. Limestone ("Loch Tay"). (Blue on map.)	
HIGHLANDS.	5. Garnetiferous mica-schists. (*g* on map.)	
	4. Green Beds. (*p* on map.)	
	3. Schistose Grits ("Ben Ledi Grits and Schists"). (*x* on map.)	
	2. Aberfoyle and Birnam Slates. (*l* on map.)	
	1. Schistose Grits ("Leny Grit"). (*x* on map.)	

Arenig ? { Grits, Black Shales, Cherts and Hornblende Schist.

Immediately to the north of the Highland boundary
fault there comes a narrow band of carbonaceous shales,
grits and cherts, which appear to have been wedged in
between the Highland schists and the Old Red Sandstone.
These rocks are supposed to belong either to the Ordo
vician or Upper Cambrian system. They enter Perthshire
to the west of Aberfoyle, from which point they can be
traced to the east of Callander at Kilmahog.

R. Garry   Conbhar   Loch Tummel

Creagan Feadaire   R. Tay

Geological Section across the Grampians from R. Garry to R. Tay

(After the Geological Survey)

1. Ben Ledi Grit.   2. Green Beds.   3. Pitlochry Schist.   4. Loch Tay Limestone.
5. Garnetiferous Mica Schist.   6. Epidiorite Sill.   7. Ben Lawers Schist.   8. Black Schist.
9. Tremolite or "Little" Limestone.   10. Blair Atholl Limestone.   11. Boulder Bed.
12. Quartzite.   13. Moine Schist.

Proceeding northwards we have first a narrow band of
schistose grit, the Leny Grit, and then the Aberfoyle and
Birnam slates. These are succeeded to the north by a
broad belt of schistose grits, which form the great mountain masses of Ben Venue, Ben Ledi, and Ben Vorlich,
and which give rise to much of the rugged scenery of the
Highland border. Succeeding these come the limestone
series of Loch Tay, followed by the garnetiferous schists,
Ben Lawers schist, Black schist, Blair Atholl limestone,
and the quartzites and quartz schists of the central Highlands. Still further to the north comes a group of schistose
rocks known as the Moine schists, whose exact geological
relationship has not yet been determined.

All these rocks have been thrown into a complicated
series of folds. One of the main axes of folding coincides
with a line running from Tyndrum along the north side
of Glen Dochart and Loch Tay, and passing through the
summit of Ben Lawers. Further to the north-east it can
be traced from Cammoch Hill across the lower part of
Strath Tummel to the Garry, and from thence eastwards
in the direction of Ben Vrackie. From this great axial
line of folding the schists have been thrown off to the
north-west and south-east in a series of minor folds. The
general structure of the ground and the relationships of
the different schist zones will best be understood by an
examination of figure, p. 37, which gives a section across
the Highlands from Glen Lyon through Ben Lawers to
the village of Comrie on the Highland border.

The geological structure of the Old Red Sandstone
area in Perthshire shows that along the southern margin

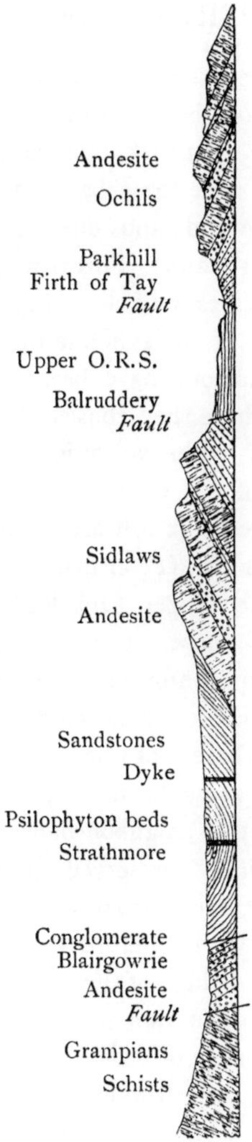

Andesite
Ochils

Parkhill
Firth of Tay
Fault

Upper O.R.S.

Balruddery
Fault

Sidlaws

Andesite

Sandstones
Dyke

Psilophyton beds
Strathmore

Conglomerate
Blairgowrie
Andesite
Fault

Grampians
Schists

Geological Section across Strathmore to the Ochils

of the Highlands there occur a massive series of conglomerates, which have been thrown down against the schists at high angles. In making a traverse towards the south-east it is found that these basal conglomerates pass into fine beds of shale and sandstone that are bent into a synclinal trough (c^1 on map). This trough or downward fold of the rocks coincides with the valley of Strathmore.

On the south side of the syncline the oldest members of the Old Red Sandstone have been exposed near the Yetts of Muckart, where they consist of coarse agglomerates and lava flows. The volcanic rocks forming the great anticlinal arch of the Ochils and Sidlaws consisting of beds of lava and volcanic ash are estimated to have a thickness of over 6000 feet (P on map).

The rocks of the Upper Old Red Sandstone rest unconformably upon those of the Lower Old Red Sandstone and pass up conformably into the Calciferous Sandstones of the Carboniferous system (c^3 on map). Along the Carse of Gowrie these rocks have been preserved in a remarkable manner, having been let down between two powerful faults. In the neighbourhood of Clashbennie they have yielded finely preserved specimens of the characteristic fishes of this formation.

A small patch of Carboniferous rocks appears in the neighbourhood of the Bridge of Earn. This is the only representative of that formation to be seen north of the Ochils. The strata consist of beds of blue-clay, sandstone and calcareous bands, and belong to the Cement-stone series lying at the base of the Carboniferous system. The presence of this outlier is of great geological interest as it points to the former wide extension of the Carboniferous formation over Perthshire, from which it has now been almost entirely removed by denudation.

The metamorphic rocks of the Highlands have been pierced by intrusions of igneous material, some of which are older and some later in time than the movements which produced the metamorphism in the schists. The earlier intrusions are represented by gneissose granites and hornblende schists (*Bg* on map), while the later consist for the most part of great masses of granite and sills and dykes of quartz-felsite (*F*, *D* and *G* on map).

Numerous dykes of dolerite cross the county in an east and west direction (*B* on map). Two of these after traversing the volcanic rocks of the Sidlaws, strike across the Old Red Sandstone rocks of Strathmore, and enter the Highland region near Glenartney, where they cut obliquely across the fault line, continuing westward by Loch Lubnaig and Loch Katrine to Loch Lomond.

Abundant evidence is to be found throughout the county of the glacial conditions that existed in Scotland in (geologically speaking) comparatively recent times. Ice-worn surfaces occur even on the highest summits of the Ochils and the Sidlaws, and the peaks of some of the Highland hills show similar striations. On the top of these glaciated rock surfaces comes the boulder clay, often reaching a considerable thickness in Strathmore. In the Highlands fine examples can be seen of the moraines formed during the later valley glaciation. These are especially well developed in the valley of the Dochart, near Killin, and on the banks of Loch Katrine between Stronachlachar and Loch Lomond. Travelled boulders are to be met with all over the region. Many boulders of Highland schist have been carried across the valley of Strathmore and deposited on the slopes and summits of the Ochils and Sidlaws. A finely laminated brick clay containing arctic shells rests on the boulder clay of the Carse of Gowrie. The arctic or sub-arctic shells found in these

Campsie Linn on the Tay
(*A dolerite dyke*)

deposits are not found living in the British seas at the present day, but exist in those of more northern latitudes such as Greenland and Spitzbergen.

The soils of the Highland region of Perthshire have been largely derived from the destruction of the crystalline schists, and generally present an arenaceous or sandy rather than an argillaceous or clayey character. As a rule they are of no great depth, and suffer greatly in dry seasons from the absence of moisture. In the Highlands, where the boulder clay exists as a soil, most of the arable farms are confined to this deposit. Over the morainic drift areas the farms are generally pastoral. The most valuable soil occurring within the Highland district is the fine alluvium to be found in the river valleys. Considerable alluvial tracts can be seen around Killin, in various parts of Glen Dochart, Strath Fillan, and in other glens in the county.

In the Lowland region of Strathmore the arenaceous element also enters largely into the composition of the soils. Usually, however, they have more peroxide of iron than the Highland soils, as well as being richer and deeper. The alluvial deposits formed by the rivers also cover much greater areas than they do in the Highlands. The flat tract lying along the valley of the Forth from Gartmore Bridge to the Bridge of Allan consists of a thick bed of stiff clay. A similar bed of clay covers by far the larger part of the Carse of Gowrie. The soils covering the sides of the Sidlaws and Ochils are rich in soda, potash and magnesia, derived from the disintegration of the volcanic rocks which form these hills.

7. Scenery and Geology.

We now pass to a brief consideration of the relationships that exist between the geological structure and the scenery of the county. It was shown, in the section dealing with the surface and general features, that the Highland area may be looked upon as a great plateau which has been dissected by the rivers flowing in a series of longitudinal and transverse valleys.

The general dead level to which the Highland hills rise is, as we learned, called by geologists a plain of marine denudation, and the only agent that could have produced such a plain is the sea. At one time, then, the sea must have cut clean across the Highland region, burying it under a great mass of its own ruins, part of which is represented by the materials that went to form the Old Red Sandstone and Carboniferous formations. The transverse valleys would have their initial direction given to them by the slope of the marine plain of denudation towards the south-east. It seems highly probable that the direction of those streams would be determined when as yet a thick covering of Old Red Sandstone rested upon the underlying schists; and when the streams reached the schists, they would continue to keep their original courses.

Water falling upon the sides of the original transverse valleys instead of following the outward slope of the plain would begin to form tributary streams which would lie parallel to the general strike of the rocks. In this way such deep longitudinal trenches as the valley of the Tay from Ballinluig to the head of Glen Dochart would be formed. The Highland section of the Tay may, then, be divided into three portions: *first*, the short transverse valley of the Fillan; *second*, the great longitudinal valley just described; and *third*, the transverse portion from

Ballinluig to Dunkeld, which is simply the southern prolongation of the great transverse valley of the Garry.

Another series of transverse streams occurs to the east of the Tay valley, the principal of these being the Ardle and the Shee, which unite to form the Ericht. After descending through the Highland schists and crossing the boundary fault, they are caught up by the Isla, which after a similar Highland course bends sharply round to the west near Alyth and flows in a longitudinal valley along the syncline of Strathmore to join the Tay near Cargill.

Turning to a consideration of the Lowland portion of the Tay valley, we find that after passing in a broad loop over Strathmore from Birnam to Perth, the river is again caught up by a longitudinal valley and carried in a north-easterly direction to the sea, between the Sidlaw and the Ochil Hills.

In attempting to account for this portion of the Tay valley, it will at first seem strange that the Tay should have selected to find its way to the sea along a ridge of volcanic rocks rather than by the synclinal trough of sandstones forming Strathmore. The reason for this will, however, be easily understood if the reader recalls the fact that a great trough fault passes along the axis of the Sidlaws and the Ochils, bringing into the centre of the arch of volcanic rocks a series of softer sandstones. This structure would play a most important part in determining the operations of the denuding forces as the soft sandstones would be more easily worn away than the volcanic rocks forming the sides of the trough, and in this manner the present valley of the Tay below Perth has been formed.

Such, then, appear to have been the main lines upon which the outstanding physiographical features of the county have been evolved. It will be seen that in few

Glen Ample
(*A valley caused by a fault*)

cases can a valley be directly traced to the occurrence of a fault or crack in the rocks. One notable exception to this is the valley of the Ample, which enters Loch Earn near its western end. The direction of this glen can be directly traced to the existence of a great fault which throws the hard grits of Ben Vorlich to the east against a series of soft schists to the west.

It has already been pointed out that the schist bands traverse the Highlands in a general north-east and south-west direction; and to the varying characters of these schists much of the picturesque scenery of the Highlands is due. The band of slates along the Highland frontier

Ben Venue
(*Showing scenic character of Ben Ledi Grits*)

forms hills of a smooth undulating character. Behind this come the massive grits of Ben Venue, Ben Ledi and Ben Vorlich; and it is the presence of these rocks that gives rise to the wild and romantic scenery of the Trossachs and the Pass of Leny, which has been so vividly described by Sir Walter Scott in *The Lady of the Lake* :

"The rocky summits, split and rent,
Form'd turret, dome, or battlement,
Or seemed fantastically set
With cupola or minaret,
Wild crests as pagod ever deck'd
Or mosque of Eastern architect."

The scenery produced by the garnetiferous schists and the Ben Lawers phyllites is often grand and imposing.

Schiehallion
(*A mountain of quartzite*)

It is typically developed along the ridge that lies to the north of Loch Tay, the rugged outlines of Creag na Caillich, Meall Garbh, and Meall nan Tarmachan corresponding to a belt of Ben Lawers phyllite resting upon a base of garnetiferous schists. The quartzites and Moine schists of north-west Perthshire frequently give rise to mountains having a more or less well developed conical outline such as Schiehallion and Ben Doireaan.

The boundary line between the Highland schists and the Old Red Sandstone is of course the great outstanding scenic feature of the county, but this has been so often referred to already as not to require any further description.

The Ochils and the Sidlaws present a low chain of round-backed swelling hills intersected here and there by defiles or passes. In the geological section it was shown that these hills consist of a thick series of lava beds bent into an anticlinal arch. On the north-west side of this arch the lava beds slope away gently to the north-west, generally presenting bold mural escarpments towards the south-east. This characteristic and often strongly marked feature can be well seen from the summit of Moncrieff Hill, near Perth. It is typically developed both in Kinnoull Hill and Dunsinane Hill.

Everywhere throughout the county the long period of glaciation has stamped with more or less distinctness its influence upon its physical features. Many of the Highland valleys are beautifully rounded and smoothed in the direction traversed by the ice; and great accumulations of morainic material make prominent features in the landscape. In the Lowland area the thick accumulations of boulder clay rise into long characteristic hummocky ridges.

8. Natural History.

Within recent years a growing importance has been attached to the geographical distribution of plants and animals. This has thrown not only a flood of light upon the past history of the earth, but it has also helped to clear up many points bearing on the relationship and origin of species.

As will readily be understood, Perthshire contains but few plants or animals that are not to be found in other parts of Britain. But it does not follow that they are to be found in all parts of the island. Thus some species have their northern limit while others have their southern, eastern, or western limit within the county.

Nor does Britain contain many that are not inhabitants of the rest of Europe. Further it will be found that the fauna and the flora of Europe are characteristic of a great region which stretches from Britain to Japan, and from the north Pole to North Africa and the Himalayas, known as the Palaearctic Region.

It is now generally believed that the greater part of the British fauna and flora reached these islands by a land connection with the Continent. From evidence, into which we cannot now enter, it is supposed that towards the close of the Ice Age the British Isles underwent a slow upheaval to a height probably corresponding to the 80 fathom line, the consequence being that the present bed of the North Sea was elevated into land, through which flowed the Rhine with the Thames, Ouse, Tay and other British rivers now entering the North Sea, as its tributaries. At this time the English Channel, St George's Channel and the Irish Sea were also land, forming a group of low-lying grounds uniting Britain and Ireland to the Continent so that the immigration of the arctic-alpine flora and fauna took place step by step across the plains from these centres of dispersion till they covered the whole of the British Isles.

Towards the close of glacial times, when the great ice sheet had passed away and only local glaciers were to be found here and there in the mountainous districts, the low grounds of Central Europe were covered by an arctic-alpine flora and fauna. With the gradual amelioration of the climate these plants and animals were forced to retreat to higher latitudes, while those inhabiting Central Europe retreated to the higher mountains, closely followed by the incoming march of the temperate species. There can scarcely be any doubt that it was this arctic-alpine flora that first covered these islands after the retreat of the glaciers.

The commonest animals in Britain at that time were the reindeer, the elk, the mammoth, the wolf and so forth. After the retiral of these northern plants and animals to higher latitudes, the country was invaded by a temperate flora which is now the prevalent type of vegetation.

It is impossible to say how long the land remained at this high level, but there is strong evidence to show that when the existing fauna and flora migrated into Britain the country was undergoing a gradual subsidence. As a result of this Ireland was first of all separated from

4—2

England, and at a later period England was separated from the Continent. The earlier separation of Ireland from Britain explains the comparative paucity of mammals and reptiles in the former country. That is, Ireland had been cut off before these animals had ceased to migrate into England.

The Highland region of Perthshire, especially Breadalbane, has long been famous to botanists because of the richness of its alpine flora. Thus on a series of mountains which stretch from Ben Laoigh north-eastwards through Meall Ghaordie and along the ridge bounding the north of Loch Tay and including such peaks as Craig na Callich, Meall nan Tarmachan, Beinn Ghlas, and highest of all Ben Lawers, and from Breadalbane north-eastwards into Clova, we find an exuberant development of alpine plants. Another tract also exceedingly rich in alpine species is from Ben Laoigh northwards by the heads of Glen Lochay and Glen Lyon and includes the following mountains: Cam Chreag, Creag Mhor, Ben Heasgarnich and others.

On the summit of Ben Lawers the schistose rocks have been weathered into a series of rock-girt pits or hollows, which form the abode of *Saxifraga cernua*—its only station in Great Britain. On the theory that it with its fellows once covered the lowlands, its solitary position here has been not inaptly called its last citadel. Step by step the northward march of the temperate flora has pushed it from the plains to the hills and from the hills to the mountains. Along the Ben Lawers ridge many other alpine species are to be found, as *Gentiana nivalis, Salix herbacea, Saussurea alpina, Erigeron alpinus* and *Dryas octopetala.*

Dryas octopetala on Ben Laoigh

It has been shown that the present distribution of the alpine flora in the Perthshire Highlands and the mountains richest in these alpine species coincide with the outcrop

of the schists known as the Ben Lawers phyllites.   The
minute structure and chemical composition of these schists
as well as the altitude that they reach form a favourable
environment for the last stand of the alpine plants.

It is calculated that the flora of Perthshire comprises
upwards of 1200 species and varieties of flowering plants,
ferns, etc., and from such a number it is difficult to single
out particular species for special mention.   Throughout
their whole length the valleys of the Tay, Tummel, Garry
and other Perthshire rivers present an exuberant and interesting flora; while the shores and waters of the numerous lochs are particularly rich in plant life.   The chain
of lochs lying between Blairgowrie and Dunkeld is perhaps
the most productive, especially in pond weeds (Naiadaceae),
the beautiful plant *Naias flexilis* occurring in several of
the lochs.

Of the 80 orders into which the trees and shrubs
of Britain are divided 19 are found in Perthshire.   The
common hawthorn, for example, is found in its wild state
as a shrub or a tree.   The crab or wild apple occurs in
hedgerows and waste places; and the mountain ash on the
seashore and on the tops of mountains as high as 2500
feet.   The common elder is most abundant in coppices
and woods.   The Scots or wych elm is also found.
Among the willows we have the brittle-twigged or crack
willow.   The aspen poplar, the common alder, the birch,
the oak, the common hazel and the Scots pine are also
plentifully distributed throughout the shire.

Coming now to the fauna, we find that 43 mammals
are recorded as occurring within the county.   Nine of

Red Deer, Glenartney

these, however, must be regarded as exceedingly scarce or practically extinct, while ten have only one or two records each. Four species of bats have been recorded, including the rare whiskered bat. Among the Insectivora the hedge-hog, the mole and the common shrew are abundant. The Carnivora are represented by the wild cat, which is, however, very infrequent; the fox, common on the mountains. The weasel and the stoat are plentiful. The badger is still found on the mountains but with a somewhat limited distribution. The otter abounds in many of the rivers and lochs. Three species of seals have been recorded from the Tay estuary. Red deer and roe deer occur in the county, the former being mostly confined to the Highlands. Among the rodents the following may be noted as natives of the shire—the squirrel, the brown rat, the common mouse, the wood mouse, the common field vole, the red field vole, the water vole, the common hare, the mountain hare, and the rabbit.

The birds of Perthshire include 228 species, of which 74 reside in the county throughout the year, and 24 for only part of the year; 34 come as summer visitors, and 25 as winter visitors; 9 are annual spring or autumn migrants, and 62 only occasional or rare visitors. It is estimated that 127 species nest in the county.

It would be impossible to give here a detailed account of the birds of the county. The higher mountains of the Highland area afford an occasional resting-place for the golden and the white-tailed eagle, while the peregrine falcon is known to nest in the shire. The ptarmigan and the snow bunting also breed on some of the higher summits. The county abounds in game birds of all kinds, especially the red grouse, the pursuit of which annually attracts sportsmen in great numbers. The capercailzie,

Hen Capercailzie on Nest

originally a native, became extinct, but was reintroduced from Norway and is now abundant.

The well-wooded glens and valleys afford a favourite

resort for warblers and small birds of all kinds. The
kingfisher, bald coot and water-hen inhabit the banks of
rivers. The oyster catcher is abundant, breeding freely
on many of the islands and banks of the Tay and
Tummel. The raven, though by no means common, is
still to be found among the mountains, while the hooded
crow abounds. The rook, known more generally as the
crow, is abundant. The jackdaw, magpie and jay, though
formerly common, are now more scarce.

Many of the lochs are particularly rich in sea birds,
ducks, geese, etc. Among the birds killed on Loch Tay
may be mentioned the osprey, wild swan, pochard, wid-
geon, tufted duck, golden-eye, scaup duck, goosander, little
grebe, great northern diver, cormorant, razor-bill, puffin,
and Leach's petrel. Others found in the immediate neighbourhood of the loch are the snowy owl, woodpecker,
Bohemian waxwing, snow bunting, brambling, crossbill,
quail and pigmy curlew.

The amphibia of Perthshire are represented by the
frog, the toad and the common newt, which are everywhere abundant. Of more restricted occurrence are the
palmated newt and the warted newt. The reptilia include three species—the common lizard, found usually
in heathery places; the slow worm, commonly frequent-
ing thick undergrowths; and the adder, which is the only
venomous reptile found in this country. It occurs in
considerable numbers in certain localities among the Highland hills.

The fish fauna of the lochs and rivers includes about
23 species. Some of the more important of these may

be briefly mentioned. The salmon occurs in all the rivers and lochs to which it can find access. The Tay has long been noted for its salmon fisheries. The sea trout is abundant in the rivers, while the common trout occurs in all the rivers and lochs, which also swarm with perch, pike and eels. The sturgeon has been taken at Perth and sprats are common in the Tay estuary. The sea lamprey is occasionally found in the Tay and has been captured as high up the river as Dunkeld. The river lamprey also occurs in the Lowland rivers and streams.

The invertebrate animals of Perthshire form such an extensive division as to preclude any possibility of dealing with them in the allotted space. Let it be sufficient to say that the land and the freshwater mollusca of the county are rich. The pearl mussel (*Unio margaritifer*) at one time occurred in the Tay and other rivers in great numbers. Recently, however, they have grown much fewer owing to the extent to which pearl fishing has been pursued. The pearls vary in colour from pure white to deep brown. *Limnaea peregra* is found in most of the rivers and also in ponds, etc. Among the land shells *Helix nemoralis* and *Helix arbustorum* are common and widely distributed. So also are *Bulimus obscurus*, *Pupa umbilicata*, and *Clausilia rugosa*, all of which are fairly common amongst stones and moss and on rocks. *Clausilia rugosa* is not uncommon on trees, which it can climb to a considerable height.

Perthshire is rich in the different orders of insects. The Rannoch district has long been famous for the northern species it has yielded, as well as for several southern species which have not been found elsewhere in Scotland. The following are some of the rare moths found in Rannoch—*Asteroscopus nubeculosis*, *Noctua sobrina* and *Nyssia lapponaria*.

The lakes and ponds of Perthshire abound in a great variety of animals belonging to the crustacean, coelenterate and protozoan divisions of the invertebrata, as has been shown by the recent discoveries made during the bathymetrical survey of these lochs.

9. Climate and Rainfall.

The principal factors in determining the climate of a country are its latitude, shape, exposure to the sea or to a particular point of the compass, its elevation above sea-level, the character of its river and valley systems, nature of its soils, and the humidity and the temperature of the air, the last two being perhaps the most important.

It would be impossible here to discuss all the principles which govern the changes of the weather. It may, however, be pointed out that the weather of the country is to a great extent influenced by *cyclones* from the Atlantic. The movements of the air may either be *cyclonic* or *anticyclonic*. Cyclones are areas of low barometric pressure with an encircling system of winds blowing spirally inwards with a direction opposite to that of the hands of a clock. Cyclonic systems usually bring to the region which they cover a large amount of cloud and rain, and may be described as bad weather systems. Anticyclones, on the other hand, are areas of high pressure from which gentle breezes blow spirally outwards, the direction of the winds being the same as the hands of a watch. This system is marked, especially towards its centre, by dry and fair weather. There are three fairly permanent pressure centres which influence the winds of Scotland throughout the year—a low pressure area south of Iceland; a high pressure area situated in the Atlantic near the Azores; and a continental area in Europe and West Asia, high in winter and low in summer. During the winter the Icelandic and the continental centres are in predominance, and give rise to a great swirl between them, which causes the wind to blow from a south-west to a north-east direction.

The direction of the prevailing winds in the neighbourhood of Perth is shown by a long series of records printed in the *Transactions of the Perthshire Society of Natural Science.* The results have been expressed in the diagrammatic form known as a wind rose (see figures on p. 62) and embody observations taken over a period of seventeen years. Along each of the eight principal points of the compass in these diagrams a distance has been marked off proportional to the percentage of days on which the wind blew in that direction. In the top diagram, which represents the winds for January, it will be seen that the prevailing winds are those from the south-west and east. The same holds good for the month of July, as appears in the second diagram, while the third diagram shows that these are the prevalent directions of the wind for the whole year.

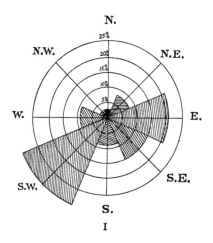

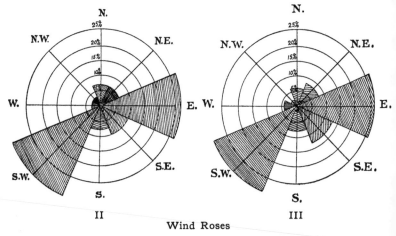

Wind Roses

(*Showing prevalent winds at Perth in January, July and throughout the year*)

We have seen that the mountainous regions of the shire lie mostly in the west and thus approximately face the rain-bringing winds from the Atlantic; but the break down of the watershed between the Firths of Clyde and Forth exposes the whole of southern Perthshire to the clouds and rains of the west. Easterly winds bring rain and unsettled weather on Gowrie, Stormont, Glenshee and Strathardle, while the weather is dry and serene in Breadalbane. It will readily be understood, however, that neither class of winds can penetrate very far into the interior without being in great part disburdened of their moisture by the mountain ranges.

The chief point that has been deduced from a large series of observations of the rainfall of Scotland is the enormous difference that exists between that of the west and that of the east. The stations along the west coast show such figures as 40, 45, and 54 inches as compared with 24, 27 and 30 inches at stations on the east coast not situated in the immediate neighbourhood of the hills. If we keep in mind that the great source of rainfall is the prevailing south-westerly winds, we easily understand that the comparatively small rainfall in such districts as the shores of the Firth of Forth and the Firth of Tay is due to the high land lying to the south-west, which robs the winds of a large proportion of their moisture in their passage across. On the other hand, the mountainous region of the West Highlands, deeply indented with arms of the sea which run in all directions from south round to west, has currents of moist air continuously poured in upon it with the result that this district has an enormously high rainfall. Thus at Loch Dhu it amounts to 82 inches, at the head of Loch Lomond to 115 inches, and at Glencroe to 128 inches. Between the extremes the amount of rainfall varies according to the physical configuration of the surface.

From the average monthly rainfall at different stations along the east and the west slope of the country for a period of years, the annual rainfall deduced from these averages is—for the whole country 44 inches, for the eastern slope 38 inches, and for the western slope 50 inches. It may be recalled that Perthshire lies almost entirely on the eastern slope, the north-western part of the watershed keeping close to the boundary line of the county.

The following table shows the gradual increase in the annual rainfall, in inches, from the east to the west of Perthshire:

Perth	32·10
Auchterarder...	39·53
Dunblane	34·49
Lanrick Castle	47·31
Loch Vennachar	58·29
Bridge of Turk	68·21
Loch Drunkie	65·13
Aberfoyle	59·54
Loch Dhu	82·73
Loch Katrine	78·42

The largest monthly rainfall occurs—in December in the mountainous districts of the interior, in January in the south-west and east of Perthshire and in the Ochil Hills.

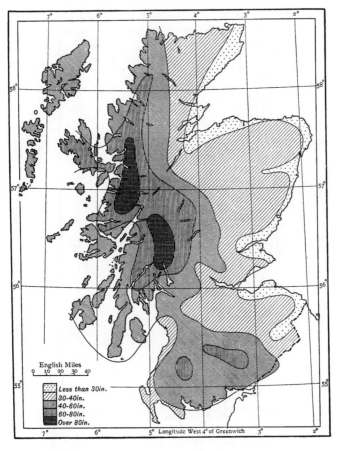

Rainfall map of Scotland. (After Dr H. R. Mill)

The diagram shows the variation of the rainfall from month to month at Perth, Lanrick Castle and Loch Dhu.

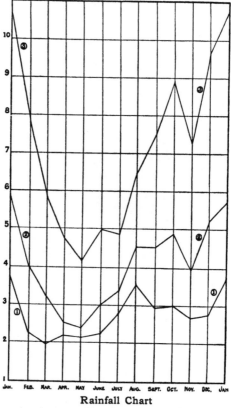

Rainfall Chart

(*Diagram showing rainfall at Perth* 1, *Lanrick Castle* 2, *and Loch Dhu* 3)

It is worthy of note that the highest rainfall at Perth never reaches the lowest recorded at Loch Dhu.

Storms of wind accompanied by great torrential down-pours of rain often lasting for several hours, and repeated over and over again during the course of a month, are

New Stream Course produced by sudden fall of rain

occasionally experienced in Highland Perthshire. Such a storm broke over Lochearnside in the month of August, 1910. The streams were swollen to such an extent that great damage was done to the crops. Roads were buried

5—2

under tons of boulders and gravel so that all traffic had
to be suspended until they were removed.   Bridges were
swept away; and in some cases water-courses of great
depth and length were cut through fields of arable land.
At Derry on the north side of Loch Earn a mountain
torrent in the short space of three hours excavated for
itself a new channel over 200 yards in length with a
breadth of from six to nine feet and a depth of from four
to six feet.   This channel coincided with a road through
a hayfield and entirely destroyed the road, while the field
was covered with tons upon tons of huge boulders.

The temperature is remarkably constant everywhere
throughout the county, averaging 47° F. for the year.
The coldest month is January (36·5° F.) and the hottest
July (59° F.).   On the whole the climate of Perthshire
may be described as mild and salubrious; and this applies
especially to the southern parts.   In the more northerly
and westerly parts, where the ground reaches a high elevation, the nature of the country makes it cold; but these
districts are also dry and healthy as they are screened from
the northern blasts by the high ridge of the Grampians.
It has been found that the death rate of a county is
determined to a considerable extent by the increase or
the decrease of cold; and as the temperature of Perthshire is fairly constant the yearly mortality varies but
little.

It has been observed in the neighbourhood of Perth
that with a north-west wind fogs never occur, very rarely
snow, and more seldom rain.   The soft heavy flakes of
snow are most common when the wind is in a south-west

direction.   Fogs prevail in the city when the wind is off the east and appear to be most common immediately after a period of dry weather.

10.   People — Race, Type, Language, Settlements, Population.

The earliest inhabitants of the British Isles of which we have any record were the men of the Stone Age. They have been divided into two periods—the Palaeolithic or Older Stone Age and the Neolithic or Newer Stone Age.   It is generally agreed that no undoubted evidence of Palaeolithic man has yet been found in Scotland, though his existence in England is shown by the numerous flint implements fabricated by him which have been found scattered over a great portion of that country.   That Neolithic man existed in Scotland is proved by the occurrence of bones, implements, weapons and other relics that belong to this period.   An ancient dug-out canoe of pine, probably of this age, was found in a brick-clay pit at Friarton near Perth.   Whence Neolithic man came and who he was is not absolutely certain.   It is generally supposed that he belonged to a non-Aryan race, of Iberian type, short-statured and long-headed people who buried their dead in chambered graves of the long-barrow form.

Long before historic time these early inhabitants of our country were pushed away to the more inaccessible and mountainous regions of the west and north by the incoming of a taller and more powerful race of a Celtic

Aryan type—the Gaels or Goidels, from whom are descended the great mass of the Gaelic-speakers who have inhabited Ireland, the Isle of Man and the north of Scotland. The Gaels were in turn displaced by a fresh wave of Aryans—the Britons or Brythons, who also belonged to the Celtic race but who spoke a different dialect.

Tacitus' narrative of Agricola's campaigns (80–85 A.D.) in North Britain gives no precise details about the tribes then inhabiting modern Perthshire. The Alexandrian geographer, Ptolemy, in the second century A.D., informs us that the region of what is now Menteith and Strathearn was occupied by part of the great tribe of the Damnonii, while to the north lay the Vacomagi. In later centuries the people of Perthshire belonged to the southern division of the Picts. In the fifth century— perhaps earlier—Teutonic invaders came from over the German Ocean, and in time penetrated the Lowland parts of Perthshire, driving the Celts to the fastnesses of the hills. That the Celts once occupied the whole of the Lowland region is shown by the fact that many of the place names are of Gaelic origin. Thus we have Auchtergaven, *uachdar-gamhauin*, "upland of the yearling cattle"; Auchterarder, *uachdar-ard-thir*, "upper highland"; and Doune, meaning "the hill." These Celts and Teutons are, in the main, the progenitors of the present-day inhabitants of Perthshire.

Up till the present day the Highland boundary line has existed as a sharp line of demarcation between the Celtic and the Teutonic race. To the north of that line

Gaelic is the vernacular tongue, to the south English is the universally spoken language. It has been estimated that about 14,124 persons or 11·55 of the population speak Gaelic.

The Scottish language originally meant the Gaelic language, but as the Teutons gradually became the dominant race the term Scottish was applied to the Anglic dialect of the Lowlands, which came from the Northern dialect of England. Latterly in Lowland Perthshire, as throughout the Lowlands generally, a form of Northern English became the vernacular.

The population of the county is but sparsely distributed. At the beginning of the nineteenth century it numbered 125,583. It reached its maximum in 1831, 142,166; and then slowly declined to 123,283 in 1901, over 2000 less than it was a century before. It has since risen slightly, to 124,339 in 1911. It may be pointed out that while there has been a considerable growth in the population of one or two of the residential villages and commercial towns, there has been a very serious fall in the rural population. This can be attributed to several causes, such as the attraction of town life, emigration to foreign countries, the growth of railways, the competition of foreign food supplies, and lastly the demands of the sportsmen, from whom the proprietor can obtain a much larger rent than he could by letting the ground to Crofters. There can be no doubt that clearances took place in different parts of Highland Perthshire towards the close of the eighteenth century and at still later periods. But it seems most likely that

even though such clearances had never occurred, the same
depletion of the Crofter population would have taken
place as a consequence of the development of the great
mineral wealth of the midland counties, towards whose

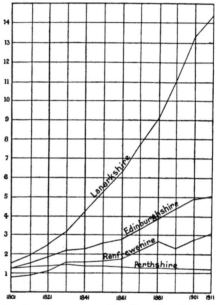

Curve showing the comparative Growth of the Population
of Perthshire, Renfrewshire, Edinburghshire, and Lanarkshire

different centres of industry the Highland population
naturally tended to gravitate. Pennant, in his *Tour
through Scotland*, made in 1769, gives us the following
description of the population and industrial conditions of
Loch Tay, with which the existing condition of things

seems to compare very unfavourably. "The north side of Loch Tay is very populous; for in sixteen square miles are seventeen hundred and eighty six souls, on the other side above twelve hundred. The country, within these

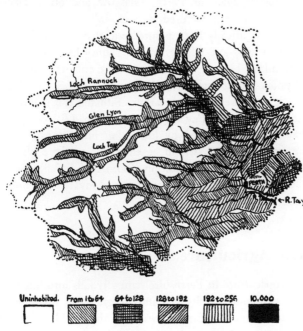

Density of Population in Perthshire (per sq. mile)

thirty years, manufactures a great deal of thread. They spin with rocks, which they do while they attend the cattle on the hills; and, at the four fairs held in the year at Kinmore, above sixteen hundred pounds worth of yarn

is sold out of Breadalbane only : which shows the increase
of industry in these parts, for less than forty years ago
there was not the least trade in this article. The yarn is
bought by persons who attend the fairs for that purpose
and sell it again at Perth, Glasgow and other places,
where it is manufactured into cloth."

It has already been pointed out that the distribution
of the towns and villages in the county has been largely
determined by its geological structure and its consequent
physiographical features. The site of the city of Perth
scarcely requires any explanation. Situated as it is near the
tidal limits of the Tay and lying on a broad flat of alluvial
haughland surrounded by low lying hills, Perth, it can
be easily imagined, would be early chosen for a place of
settlement. Note the position of Dunkeld, Crieff, Comrie,
Callander and Aberfoyle at the gateways to the mountains.

11. Agriculture.

Agriculture in Perthshire naturally falls into line with
the two great geological and geographical divisions of
Highland and Lowland. North of the Highland boundary line the valleys are more or less restricted, being
hemmed in by lofty hills, and the country is mainly given
over to the pursuits of the artist, tourist, and the health
and pleasure seeker. The steepness of the slopes prevents
the formation of soil, and farming in its different varieties
is of secondary importance. Crops are cultivated here

and there where the valleys widen sufficiently to admit the existence of a tract of level ground, but such are very limited; and no part of the Highland area can properly be termed agricultural land. In the Highlands the main valleys would be the first to attract the population, and for this reason the valley of the Tay, which contains by far the largest amount of alluvial and arable land, is the most extensively cultivated. Another factor of considerable importance in attracting the population would be the southerly aspect of the arable land; and hence we usually find the largest number of farms and crofts on the northern and north-eastern sides of the valleys. The nature of the subsoil has also played an important part in determining the distribution of the cultivated ground in the Highland valleys. Thus it can generally be shown that over the morainic areas the farms are pastoral and give support to only a limited number of people; while in those areas where the boulder clay appears, the farms are either arable or mixed arable, giving support to a much larger number of people.

The two great agricultural areas of Perthshire are the level expanses of Strathmore and the Carse of Gowrie. These make a striking contrast to the Highlands, being almost entirely devoted to agricultural purposes. They are everywhere covered with large and thriving farms and orchards, which indicate the great depth and fertility of the soil. It is estimated that only about one-fifth of the entire area of the county is under cultivation, the rest being occupied by pasture, woods and deer forests. Extensive tracts of moorland along the northern margin of

Strathmore have been reclaimed while others have been greatly enriched by the draining and special manuring of the soils and by the careful rotation of the crops. The following figures from the Government Agricultural Statistics give the acreage devoted to the different cereals during 1909—wheat 5341 acres, barley 10,602 acres, and oats 65,662 acres. Two-thirds of the area devoted to green crops is occupied by turnips, the rest by potatoes. One-third of the total area is permanent pasturage, and 930,000 acres hill pasturage. The arable land is principally confined to the drier regions of the east and southeast, where the soil is for the most part fertile. Large stretches of Tayside and the upper districts of Menteith are dotted over with orchards, their quick soil being particularly suitable for the growth of apples. The number of holdings in the shire is somewhat above 5000, the majority being under 50 acres each. They are situated mostly in the Highland valleys and in the neighbourhood of villages and small towns.

The great variety of the Perthshire pastures enables them to support a corresponding diversity of stock. About Perth, the Bridge of Earn and the Carse of Gowrie, the Angus and Fife breeds of cattle prevail. In the Highlands the Argyllshire breed is most common. The Lanarkshire breed is found in Menteith, while the Ayrshire and Galloway breeds are found at various parts throughout the county. Black cattle from Devonshire, Lancashire, Guernsey, and the East Indies have been introduced and have been blended with the other breeds.

Next to Argyllshire, Perthshire still carries the heaviest

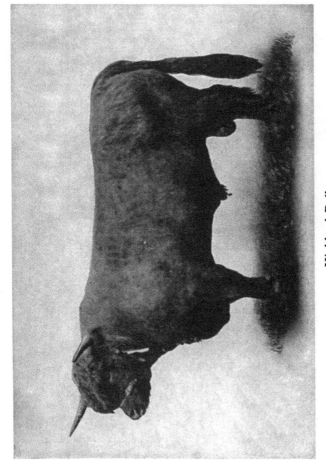

Highland Bull

flocks in Scotland. Formerly the sheep were of the white-
faced stock, which required to be housed every night
during winter. The white-faced sheep have now been
almost entirely ousted by the more hardy black-faced
sheep, either pure or cross. Cheviots, South Downs and
Leicesters are also common on the lower runs. Goats at
one time were fairly numerous throughout the county,
but they have almost everywhere given place to sheep
and tillage. Poultry and swine are common everywhere.
Dovecots occur in the neighbourhood of Perth, Coupar,
and the Carse of Gowrie.

The following table shows the average number of
live stock in four of the Scottish counties, including
Perth.

	HORSES	CATTLE	SHEEP	PIGS
Perth	13,000	72,000	690,000	8000
Ayr	10,000	100,000	380,000	14,000
Dumfries	8,000	65,000	580,000	9000
Edinburgh	4,500	19,000	187,000	9000

In Perthshire most of the horses are Clydesdales, which
are the only horses used for agricultual purposes.

At one time the whole of the county was covered by
dense forests, the remains of which are still found in the
Black Forest of Rannoch and elsewhere. The former
extent of these forests is clearly shown by the tree trunks
that are even yet dug out of the soil. Unfortunately
during feudal times these ancient forests were sadly
diminished, the common people supposing that they were
inimical to the production of food, while the barons do
not seem to have been sufficiently enlightened to stop the

work of destruction. Large numbers of trees were also cut down for fuel. In this way the low grounds were gradually divested of cover. It was this nakedness of the land that elicited from Burns *The Humble Petition of Bruar Water*, addressed to the noble Duke of Atholl, one verse of which runs thus:

> "Let lofty firs and ashes cool
>   My lowly banks o'erspread,
> And view, deep bending in the pool,
>   Their shadows' watery bed;
> Let fragrant birks, in woodbines drest,
>   My craggy cliffs adorn;
> And, for the little songster's nest,
>   The close embowering thorn."

When the late Duke of Atholl began the afforestation of his estate, he had only 1000 acres of wood on his extensive property. In 1812 Perth had 203,889 acres of woodland, which was the largest acreage of any county in Scotland at that time. In 1871 it had fallen to 83,525 acres. In 1881 it stood at 94,568 acres. The returns of 1891 show that in extent of woodlands Perth with 93,233 acres had fallen to the third place among the Scottish counties, Aberdeenshire coming first with 108,858 acres, Inverness next with 98,738 acres. Perthshire could easily afford to plant an additional 200,000 acres, and it is satisfactory to know that within recent years a movement has been set on foot to achieve this end. It was estimated by Mr Hunter that in the year 1883 the value of the woods in the county was about three and a half million pounds sterling.

Falls of Bruar

Reference has been made to the great diminution of the Crofter population in Perthshire within the last century. The wholesale clearance of tenants from their Crofts was maintained by them to be a violation of an implied security of tenure and it led in the past to a great deal of agitation by the Crofters for the purpose of securing a consideration of their grievances. The Crofters Act of 1886, and subsequent amending acts, have greatly improved the condition of the Crofters. It may be noted, however, that when that act was passed Perthshire and Aberdeenshire were exempted as they were supposed to be under more favourable economic conditions than other districts, and accordingly not to require the benefits of special legislation.

12. Industries and Manufactures.

With the exception of the citizens of Perth, the inhabitants of the county from time immemorial have been engaged mostly in agricultural and pastoral pursuits. Other occupations are mining and quarrying; and here we have to consider textile and other industries. The linen trade, though long established in the county, has never become of much importance compared with other parts of Scotland. The chief centres in Perthshire are Perth, Coupar-Angus and Blairgowrie. The cotton industry, though at one time in a flourishing state, has now passed into a condition of decadence. The principal mills are situated at Deanston and Stanley. Woollen manufactories on a small scale have been established in several towns and villages. Tweeds are manufactured at Pitlochry and Killin; tartans and galas at Auchterarder, Crieff, Dunblane, Kincardine and Burnfoot in Glendevon. Several bleachfields in the neighbourhood of Perth have existed for a considerable time.

During the latter half of the eighteenth century the city of Perth gave promise of becoming a great commercial centre, but this was soon blighted. It appears that the town reached the climax of its prosperity in 1794 or 1795. At that time linen was the staple manufacture, and it is estimated that about 1500 looms were then engaged in the town and suburbs in the manufacture of linen and cotton, the annual value of which was about £100,000. About this time a number of enterprising Perth manufacturers established bleachfields and printworks; while the making of boots and shoes in the city was also of considerable importance. These articles to the value of about £8000 were shipped yearly to London. Throughout the city there were various tanneries, which annually prepared from 8000 to 10,000 hides and about 500 dozen calf skins. Such was the state of trade in Perth towards the close of the eighteenth century. The introduction of cotton was the principal cause of the decline of the linen trade. The Perth manufacturers stuck to the linen trade, and when cotton goods came into general use, they retired from business altogether.

During the latter half of the eighteenth century there were in the neighbourhood of Perth three mills for the manufacture of paper. About the same period too the ancient fraternity of the glovers and skinners were doing a prosperous business: 30,000 sheep skins and lamb skins were dressed, and from 2000 to 3000 dozen pairs of gloves were made annually. The printing of books was also carried on, the yearly output being from 20,000 to 30,000 volumes.

Arkwright Mills, Stanley

Perth cannot now be looked upon as a great manufacturing centre. Bleachfields and printworks have passed out of the hands of her citizens; and the glove trade is entirely a thing of the past. The tanneries have greatly diminished in importance, and the volumes that were formerly issued by the thousands are now but seldom seen.

At the present day Perth may be looked upon as

6—2

a great central mercantile dépôt for the supply of the
necessaries, conveniences and luxuries of life to the surrounding districts. It is famous for its cattle markets and
for its dye works, which within recent years have assumed
considerable dimensions. The manufacture of jute and
linen is still carried on at the Perth Jute Works and the
Wallace Linen Works. Of the other industries of the

Pullar's Dye Works, Perth

city the following may be mentioned—the manufacture
of glass, ink, floorcloths, ropes and twine, bricks and
chemicals. Several grain mills give employment to a
number of the inhabitants. There are also breweries
and distilleries, ironworks and foundries.

At one time the merchants of Perth carried on an
extensive trade in their own ships with the Netherlands.
Germans and Flemings at an early period frequented the

city, and many of them settled in it. At that time ships went up the Tay as far as the Palace of Scone, for in one of the charters of the Abbey we find that Alexander I, having granted to the monastery the customs of ships coming to Scone, gave liberty to English ships to trade there, and promised them protection on paying customs to the monks. In 1830 shipbuilding began to be carried on in Perth; and some years afterwards the first iron steam vessel built on the east side of Scotland was launched from a Perth yard. But this industry has now dwindled.

13. Mines and Minerals.

The metalliferous mines of the county that have been worked to any extent are practically confined to Breadalbane, and to that part of it drained by the upper reaches of the Tay. None of these mines, however, are worked at present, though mining for lead was carried on at Tyndrum up to the year 1862.

At Tyndrum the Ben Lawers phyllites have been faulted against a series of quartzose rocks, the fault trending in a north-easterly and south-westerly direction, and running across the Strathfillan and Coninish valleys. A belt of high ground terminating to the east in a mountain called the Sron-nan-Colan is the ridge in which the lead workings are situated, and chiefly in the height named.

The levels of the workings have been driven into the Sron-nan-Colan to catch the vein as it passes through the hill. The veinstone is pure white quartz in the hard vein, and breccia made up of quartzite and mica-schist in the clay or soft vein. The principal ore is argentiferous galena (sulphide of lead with silver), zincblende, cobalt; copper and iron pyrites are also found.

The vein at Tyndrum was discovered by accident in 1741. At that time the Breadalbane minerals were leased to Sir Robert Clifton, who between that year and 1745 raised 1697 tons of lead ore. For the next 15 years the mine was worked by the Mine Adventurers of England, who extracted from it 2046 tons of ore. Between 1760 and 1762 the Ripon Company mined 330 tons of ore; and between 1762 and 1768 Messrs Richardson and Paton mined 942 tons. In 1768 the Scots Mining Company acquired the lease and began the working of the mines in a more vigorous and systematic manner. Previous to this the ore had been carried by way of Loch Lomond to Glasgow, to be shipped to the south. But the new Company erected smelting works about a mile east of the mine, and between 1768 and 1790 extracted 1678 tons of lead from 3685 tons of ore. Up to the year 1858 mining was carried on only intermittently. But in that year the late Marquis of Breadalbane took the mine into his own hands and worked it till his death in 1862.

On the southern side of Glen Lochay about three miles from the foot of the glen a bed of serpentine crops out and was mined—but on a very limited scale—for chrome iron ore by the late Marquis of Breadalbane. The serpentine is of a dark colour, mottled with lighter shades.

The chrome iron ore is disseminated through the serpentine in grains and with it are associated steatite (soapstone), chrysotile, etc.

At Tomnadashan, a hamlet situated on the south side of Loch Tay about nine miles from Killin, the mica-schists have been penetrated by a boss of granite and diorite, in which a number of metalliferous ores were discovered. Large cave-like openings have been made in the face of the hill by the removal of the rock containing the ore. At the bottom of these openings a level may be seen driven into the side of the hill. This was constructed under the supposition that the ores were concentrated in a vein, but as such was not the case, no vein was ever reached.

The chief ores are copper pyrites (chalcopyrite) and grey copper (tetrahedrite). The ore is disseminated through the igneous rock in irregular masses so that its working must always be more or less precarious. When stamped and dressed, the ore was shown, on analysis, to contain very little copper—3·58 per cent. to 30·28 per cent. sulphur. At a spot named Corrie Buie on the south side of Loch Tay argentiferous galena veins have been worked to a limited extent. Two small lumps of native gold were discovered in the quartz as it was being crushed under the hammer.

Silver, copper, lead, and cobalt have been found in association with the volcanic rocks of the Ochils, the veinstone being usually barytes. Barytes has also been worked in the Old Red Sandstone rocks to the west of Aberfoyle.

Roofing-slates have been quarried at different points along the Highland border. Many of the old quarries have now been abandoned, but they are still worked at Aberfoyle, Birnam and Logiealmond. The chief varieties are of a pale greenish-blue, and those considered still more valuable and durable of a purplish-blue or indigo colour. The slate usually contains iron-pyrites in cubes, commonly known as slate diamonds, and, occasionally near Dunkeld, specular iron-ore.

The Old Red Sandstone rocks of the valley of Strathmore have yielded good building-stones. Along the Highland border, as at Aberfoyle and Callander, the conglomerates have been very largely used for building. Perth has been almost entirely built out of sandstones from Burghmuir and other places in the neighbourhood. The long dykes of dolerite which traverse the county from end to end, and the sills of felsite and other igneous rocks which occur in association with the Highland schists, have been extensively quarried for road metal.

Very large deposits of peat are to be met with both in the Highlands and in Strathmore. They have been formed by the annual growth and decay of vegetable matter. The mosses are the most important peat-forming plants and chiefly belong to the genus *Sphagnum*. Up till recent times peat was the principal fuel in the Highlands. But the increased facilities for the transit of coal from the south have led to the gradual diminution of its use.

None of the springs which occur in Perthshire, with the exception of those at Pitcaithly, can be considered as remarkable. The mineral wells at Pitcaithly, five in

Aberfoyle Slate Quarries

number, are believed to be amongst the oldest natural
medicinal waters in the country, and are esteemed as
highly beneficial in certain complaints. Those grounds
where extensive beds of gravel rest on compact even
boulder clay usually yield the most abundant and pure
supplies of water. Chalybeate springs are occasionally
found. In the Highland area they appear in association
with the black schist. One below Blackcroft in the Pass
of Lyon has a considerable local reputation.

14. Fisheries and Fishing Stations.

The Tay and its affluents with their varied tributaries
afford a splendid breeding ground for the salmon. Along
the whole course of the Tay from the sea to the rivers
Dochart and Lochay, salmon are more or less abundant.
Loch Tay is much frequented by anglers, and large fish
are often caught. Salmon weighing 48 lbs. have been
caught with the rod; and a salmon weighing 35 lbs. is
by no means an uncommon fish. In fact it seems to be
an exception to find any fish under 18 or 20 lbs. The
Earn, the Lyon, the Tummel and the Isla are also good
salmon rivers.

The commercial fisheries of the Tay are chiefly
situated between Perth and Newburgh, on some six or
seven miles of the river. The fish are caught by the aid
of the net and coble. Many persons find employment
in the working of the different "shots," as the fishing
stations are named; and a considerable sum is annually

paid in wages.   The salmon fisheries of the Tay are
owned by various noblemen, gentlemen and corporations,
and have yielded within the last twenty years a gross
annual rent varying from £17,819 to £23,715.   It has
been estimated that the number of salmon and grilse
caught in the Tay range from 75,000 to 100,000 a year.

For a period of over 25 years the salmon hatchery
at Stormontfield supplied the river Tay with young
fish, the fry of the salmon (parr and smolts) being

Salmon, 55 lbs.

reared on what is known as the " piscicultural system."
The ova are laid down in boxes filled with gravel, over
which a stream of water is allowed to pass.   In a period
varying from three to four months the eggs are hatched.
The usual time for the hatching of salmon eggs in our
northern rivers is 130 days or between four and five
months.   This varies, however, according to the openness
or the severity of the season.   Since the closing of the
ponds at Stormontfield the breeding of salmon has been
carried on at Dupplin on the river Earn.

It is interesting to note that the natives of Perth have long recognised the necessity for, and displayed great activity in, the preservation of the salmon fry, as is shown by the following enactment: "That all cruves and zaires set in fresh water, quhair the sea fillis and ebbis, the quhilk destroyis the frie of all fisches, be destroyed and put awaie for euer mair ; not againe standing ony priuiledge and freedome given in the contrarie, under the paine of ane hundreth schillinges. And they that hes cruves in fresh waters, that they gar keepe the lawes annents Satterdaies stop : and suffer them not to stande in forbidden time, under the said paine. And that ilk heck of the foresaidis cruves be three inch wide, as the aulde statute requiris."

For a long period of time the Tay from Perth upwards was recognised as the principal seat of the pearl-fishery in Great Britain. In consequence, however, of the great destruction of the mussels by fishers the number of pearls obtained has gradually diminished. It has been estimated that only one pearl is found in every hundred shells opened, and only one in every hundred pearls is of any use for ornamental purposes. It will be manifest that pearl-fishing cannot be considered as a very lucrative business. Between 1761 and 1764, pearls to the value of £10,000 were sent from the Tay to London. This will serve to show how greatly the industry has diminished in value.

15. History of the County.

The history of the county centres to a large extent round the city of Perth. The site of the battle of Mons Graupius, in which Agricola defeated Galgacus the Caledonian general in 84 A.D., is a matter of much dispute and does not seem as if it could ever be definitely settled. Many of the leading authorities have placed the scene of the battle in Perthshire—some at Dalginross, others at Ardoch, others again at the peninsula formed by the junction of the Isla and the Tay.

How often and for how long the Romans, after Agricola's days, made campaigns and occupied strongholds in Perthshire, is as yet buried in obscurity. In later times when the county formed part of the kingdom of the southern Picts, two of their chief towns or capitals were Abernethy—as early as the sixth century—and Forteviot. Scone also became sacred as the place of coronation for the kings of the Scots. In the ninth century the centre of Celtic Christianity was transferred from Iona to Dunkeld—an event of deep significance in the consolidation of the kingdom.

According to Hector Boece the village of Luncarty situated about four miles to the north-west of Perth was the scene of the decisive overthrow of the Danes by Kenneth II.

Perth figures conspicuously during the War of Independence. The renowned champion of freedom, Sir William Wallace, was often at Perth, though exactly how often it is impossible to say. In 1297 he effected the capture of Perth then held by the English. It became one of his headquarters, and consequently, after his execution, the city was appointed to receive for a spectacle one portion of his dismembered body. Edward I was also repeatedly at Perth, and in 1296 when returning from the north visited Scone and carried away with him the records of the kingdom and the sacred stone on which the Scottish monarchs sat at their coronation.

In the year 1306 Bruce was crowned King of Scotland at Scone and shortly afterwards he made his appearance at Perth challenging, as Barbour tells, the English governor, the Earl of Pembroke, but the Earl declined the challenge, saying that the day was too far spent. He promised, however, to fight on the following day. Bruce retired with his army to Methven Wood, where Pembroke surprised him. A short but bloody battle ensued in which the Scots were routed. Bruce with the remains of his army sought safety in the Highlands. In 1311 he returned to Perth and, after besieging it in vain for six weeks, resolved to try stratagem. He retired as if he were preparing to abandon the siege, but returned during the night with a body of picked men, who waded across the ditch up to the chin in water and scaled the walls. The town was instantly taken.

The Barons who, siding with the English during the reign of Bruce, lost their lands and retired to England, descended on Scotland in 1332 and defeated the Scots at Dupplin Moor. Then their leader, Edward Balliol, was crowned at Scone.

When invading Scotland, Edward III several times visited Perth.

In October, 1396, the North Inch of Perth was thronged with spectators viewing a strange tournament. The King was there with his court; churchmen, nobles, commoners had all gathered. Two clans, usually but not certainly called Chattan and Kay, had for many a year waged war with each other; and now the quarrel was to be fought out by thirty men a-side, armed with axe and sword and knife. When the signal to close was about to be given, one of Clan Chattan (or Clan Kay, for accounts vary) was found to have deserted. For half a French crown Hal o' the Wynd, armourer and skilled swordsman, took the vacant place. A stubborn and bloody contest followed. Of the sixty combatants only twelve survived—one on the one side; on the other, eleven including the valiant substitute, whom from his bandy legs the Highlanders nicknamed Gow Chrom, "the crooked smith." It was his prowess with his two-handed sword that chiefly won victory for his side. Which that was, tradition says, he could not tell; for, when questioned after the fight, he replied that he fought for his own hand. Scott makes skilful use of the clan battle in *The Fair Maid of Perth*.

In 1407 Perth was the scene of the burning of the first Lollard martyr, James Resby, who, according to Bower, was an English priest of the school of Wycliffe. Resby had been particularly active in spreading Wycliffite doctrines.

It was at Perth that James I was murdered in 1437.

The court occupied the Blackfriars Monastery and there
the assassination took place. James had made himself
obnoxious to the lords by his arbitrary dealings with
them. On the evening of the 20th February, that arch-
conspirator, Sir Robert Graham, along with a number of
retainers, broke into the royal apartments, where the
King was chatting with the Queen and her ladies. The
bar had previously been removed from the door and the
windows of the room had been securely fastened. The
ladies could do but little to assist the King; but it is said
that one of them thrust her arm into the place of the
missing bar. The courageous deed has thus been described
by D. G. Rossetti in his *King's Tragedy*:

> "Like iron felt my arm as through
> 　The staple I made it pass—
> Alack it was flesh and bone—no more!
> 'Twas Catherine Douglas sprang to the door
> 　But I fell back Kate Barlass."

The King retired to a vault below the room, where he
was followed by the conspirators. James made a stout
resistance but was overpowered and fell with sixteen
wounds in his breast. Within a month the chief conspirators were arrested and put to death. After this
event Perth ceased to be a residence of royalty.

On the 11th of May, 1559, Knox preached, in
St John's Church, Perth, a vehement sermon against the
Mass. His hearers had not left the building when a
priest began to celebrate Mass. A youth spoke irreve-
rently of this, and the priest struck him. The boy aimed
a stone at the priest but broke an image instead. This

was like fire to gunpowder, and the "rascal multitude"
—so Knox terms them—smashed the ornaments and
furniture of the church.   Not satisfied with this, they
destroyed the Franciscan, Dominican, and Carthusian
monasteries, leaving only the bare walls.

The next important event in the history of Perth,

Gowrie House in 1805

known as the Gowrie conspiracy, took place in the year
1600.   James VI was invited to Gowrie House under
the pretext that it contained a mysterious captive with a
pot of gold.   An attempt was made to secure the King,
who gave the alarm, and his attendants rushing in slew
the Ruthvens—the Earl of Gowrie and the Master of

Ruthven. It has been asserted that this was a plot by James to ruin Gowrie and his brother, but the whole event is wrapt in mystery.

The first battle between the Marquis of Montrose and the Covenanters took place at Tibbermore on the 1st September, 1644. The Royalists won an easy victory

Gathering Stone, Dunblane

at a comparatively slight loss to themselves, and captured all the artillery and baggage of the Covenanters. From the field of victory Montrose proceeded to Perth, which next day opened its gates.

In the Pass of Killiecrankie, most picturesque of Scottish battlefields, the engagement took place which decided the fate of the Jacobite party in 1689. Over the hills came Graham of Claverhouse, Viscount Dundee, with about 2200 Highlanders and about 300 Irish recruits. General Mackay was sent north to quell the insurrection. The opposing forces met at the head of the Pass. The Highlanders reserved their fire till close on the enemy and then, throwing away their muskets, rushed on with axe and claymore, driving the royal troops into the valley below. A general panic seized them and they fled down the valley in complete disorder. Dundee was killed by a bullet and died with the notes of victory in his ear.

The last battle on Perthshire soil that we have to record was that of Sheriffmuir, fought 13th November, 1715, on the north-west side of the Ochils. The Duke of Argyll commanded the Royalist forces and the Earl of Mar those of James the Old Pretender. Both sides claimed the victory, and an old Jacobite song thus humorously hits off the combat:

"There's some say that we wan,
    And some say that they wan,
    And some say that nane wan at a', man:
    But ae thing I'm sure,
    That at Sheriffmuir
    A battle there was that I saw, man:
    And we ran and they ran, and they ran and we ran.
    And we ran and they ran awa, man."

The battle, however, checked the advance of Mar's Highlanders, and spelled disaster to the Jacobite cause. The illustration on p. 98 shows the "Gathering Stone of the Clans" on which the Highlanders are said to have whetted their dirks and claymores.

The county of Perth figures prominently in the annals of the rebellion of 1745. Charles Edward entered the county town on the 4th September. At the Cross he proclaimed his father King of Scotland and himself Regent. Charles remained in Perth for a week, drilling his troops on the North Inch.

Among the recent events connected with Perth the Meal Mobs of the latter half of the eighteenth century may simply be mentioned. What we have said confirms our original statement that the history of the county has centred to a large extent round the "Fair City." The events that were taking place in the other parts of the shire while Perth was passing through such stirring times, consisted chiefly of obscure feuds between the Highland clans; and throughout the Perthshire Highlands there are many minor battlefields marking the spots where dark and terrible deeds have been enacted. One of the most prominent and picturesque figures in the history of these Highland raids and feuds was Rob Roy, immortalised by Sir Walter Scott in the novel of that name. Macgregor Campbell or Rob Roy was in turn cattle-dealer, drover, and thief. He was involved in a dispute with the Duke of Montrose, from whose factor, Graham of Killearn, Rob Roy seized the rents paid by the tenants at Chapel Errock. Rob Roy pretended to side with Mar in the rising of 1715 and made preparations for a raid on the Lowlands by Loch Lomond side. His neutrality, however, at Sheriffmuir seems to indicate that the members of his clan who followed him were bent on obtaining booty either off one side or the other.

16. Antiquities.

To the antiquary Perthshire is full of much of the deepest archaeological interest. Objects belonging to prehistoric, Roman, Celtic and later periods are scattered over the whole county. The remains of the Stone Age have been found at altitudes varying from sea-level to near the summits of some of the Highland mountains. Stone axes have been recorded from Aberfeldy, Rattray, and

Stone Axe, found in Perthshire

other parts. Hammer-heads have been found at Dunning, Pitlochry ; and one was found along with a food vessel in an interment at Doune (see fig. above). At Perth a curious stone knife, or dagger, was found lying beside a stone cist. It is formed of a piece of mica-schist and its natural shape has been adapted to form a rude but efficient weapon. Beside the common type of axe-heads which were attached to their wooden handles by a thong, there have also been found axe-heads through which a hole has been drilled for the insertion of the shaft, some of which appear to have been used as battle-axes. A beautiful specimen of this type was found in the Tay at Mugdrum Island. Probably the earliest record that we have of man in Perthshire is a "dug-out" canoe which was discovered underneath the brick-clay at the Friarton below Perth. This canoe is supposed to belong to the earlier stage of the Neolithic Period. It shows that man occupied this district before the formation of the Carse clays and before

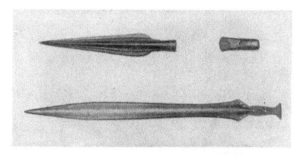

Bronze Spear and Sword from Blairgowrie, and Axe from Comrie ($\frac{1}{8}$ *actual size*)

the sea had risen to the 45 feet level. At that time the estuary reached much further inland, covering the present site of Perth to a height of 30 feet; and primitive man must have been able to paddle his canoe across what are now the streets of Perth, the river being tidal up to and beyond Stanley.

The second stage or Bronze Age shows that man had made considerable progress in handicraft, culture and civilisation, having now become acquainted with the use of metals. Numerous relics of the Bronze Age have been preserved in the county. They include flat and flanged axe-heads, from Abernethy and Perth; a fine blade and spear-head from Blairgowrie (see fig. on p. 102); rapier-shaped blades from Ardoch; and knives and daggers from Blair Drummond, Drumlanrick and Pitcaithly. Personal orna-ments such as bracelets, torques, etc., have been collected at different localities in the county. Few bronze sickles have been found in Scotland; a fine specimen, however, has been obtained from the Tay near Errol. Throughout the shire there occur many stone circles, some of them being in a wonderfully perfect state of preservation. These are generally known as Druid Temples. But they have no claim to this distinction. They are now believed to have been associated with the burial customs of the Bronze Age. Fine examples of these can be seen at Dull, Killin, Crieff, Blairgowrie and Blackford.

There is evidence that iron had been used in Britain before the advent of the Romans; and it is to the Iron Age that the great hill forts belong, which are found in such large numbers throughout the country. Some fine examples of them occur in Perth, notably the one dis-covered at Coldoch in 1870. Those windowless and roofless drystone erections have been considered by some as the immediate predecessors of the later castle. These structures seem to have been designed as retreats in time of danger for non-combatants and cattle. The great interest of the Coldoch hill fort is, that it is one of three found to the south of the Caledonian Canal. A stone fort at Abernethy, which was recently explored, yielded portions of iron implements, a bronze spiral finger-ring, fibula, bracelets, rings of jet or lignite, and a polished stone axe. A similar finger-ring was found in the hill fort of Dunsinane. The vitrified forts belong to the same period in time.

Between Blairgowrie and Meikleour may be seen the Cleaven Dyke, which runs in a straight line for 2000 yards in a north-west and south-east direction. It is twelve

Roman Camp, Ardoch

yards wide and two yards high. On each side is a level border protected by a ditch. The total width of this defensive work is 58 yards. It is supposed that this dyke was erected by the Picts as a defence against the Romans, and when it fell into the hands of the latter, they added a camp, of which traces can still be found.

Considerable diversity of opinion exists as to the many so-called Roman remains found in Perthshire, but there can be no dubiety as to the origin of the camp at Ardoch, the largest and most perfect of the kind in Scotland. It measures about 140 yards by 125 internally, and is of a rectangular shape. It is protected by a series of parallel ramparts and ditches, which are arranged in two rows on the side of the river Knaik and five on the land side. The positions of the praetorium and gateways are still quite easily distinguishable. The traces of Caledonian camps and hill forts seen in this neighbourhood indicate that at this point of their advance the Romans encountered a vigorous resistance. Some authorities place the scene of the great battle of Mons Graupius in the neighbourhood of Ardoch. Another important Roman station was that of Inchtuthil (in the parish of Caputh), a tract of land on the river Tay. It rises with steep ascent some 50 feet above the level of the neighbouring plain, and is a strong strategic position. At its north-east corner there was a Roman camp 500 yards square, whose stone walls, $9\frac{1}{2}$ feet thick, have now been reduced to the level of the surrounding ground. To the south-east of the camp there were two tumuli and a redoubt, the site of which is now marked by a group of trees. Besides these two camps there are others at Fendoch, Dalginross, Fortingall and Dunkeld. Throughout the country there are also various stretches of Roman road which, as in the policies of Gask, can be traced with a greater or lesser degree of accuracy. Here and there casual relics of the Romans have been found, such as tools and weapons of iron ; personal ornaments, including brooches and studs ; coins ; and fragments of pottery.

Cup- and ring-marked sculpturings, sometimes on separate boulders, but often on the native rocks, occur at different localities in the shire, as at Killin, Lochearnhead and Glendelvnie near Caputh. Though the exact age of these sculpturings is not known, yet there can be no doubt that they are of extreme antiquity. The exact object of these cup and ring markings has excited much curiosity and speculation amongst archaeologists. All that can be said of them is, that their origin and significance have been lost in the dim and distant past. Even tradition has nothing to say regarding them.

At Meigle, Dunfallandy, and Rossie Priory there can be seen a number of beautiful examples of Pictish monuments. These have been divided into three classes. The first and oldest consist of unshaped upright boulders, upon which have been incised certain mysterious and hitherto unexplained symbols. The second class also usually stand erect, and bear similar symbols, but accompanied by richly decorated Celtic crosses. The stones of the third and latest class are found in a recumbent position. The elaborate decoration of the second type is present, but the mysterious symbols are now wanting. The large group of these stones at Meigle—32 in all—have been collected and placed in a building with a view to their preservation. They belong to the second and third classes just described. The great Cross Slab of Meigle stands about eight feet high. The obverse shows a boldly executed Celtic cross, and the figures of a man and a beast—probably representing the story of Jonah. The reverse shows a hunting scene; Daniel surrounded by

Celtic Cross, Glencarse

Round Tower, Abernethy

lions ; a centaur, symbolic of the man-animal or the conflict between flesh and spirit. These monuments are unquestionably of Pictish origin. J. Romilly Allen considers that the larger number of them are older than 1100 A.D. The illustration on p. 107 shows the Celtic Cross at St Madoes, Glencarse, which is an elaborately sculptured stone.

Another important Perthshire antiquity is the round tower of Abernethy, which closely resembles a similar structure at Brechin. These are the only two examples of round towers in this country, though some 76 are known to exist in Ireland, having all the characteristic features of the Scottish specimens. The Abernethy tower is 74 feet high, and unlike some of the Irish ones, which are composed of rough rubble, it is built of carefully-hewn square stones. A popular idea attributed these structures to the Picts, but they are now known to have been built by early Christian architects as watch-towers, some of which have been afterwards converted into belfries.

17. Architecture—(a) Ecclesiastical.

Before the eighth century there was probably no ecclesiastical architecture of any consequence in Scotland. Such buildings as did exist were similar to those of Ireland, where the arch seems to have come into use in the ninth century, when it must have been of the simplest and rudest type. Architecture connected with church building really began about the tenth century, when the round towers first appeared.

It is usual to divide architecture between the eleventh and the sixteenth centuries into certain periods or styles, which are not arbitrary but represent distinct historical periods characterised by particular features. It must not be supposed that the change from one style to another was suddenly accomplished: as a matter of fact it usually took about half a century to effect the transition.

The Norman Style was introduced into Scotland in the twelfth century. It can easily be recognised by its simple and massive forms and semi-circular arches. The exterior is generally plain. The principal ornamentation is connected with the doorways, which are often deeply recessed, the arch mouldings being decorated with chevron or zigzag carving. The tower of Dunblane Cathedral is an example of the Norman style in Scotland.

The Norman style of architecture prevailed in Scotland for some time after the close of the twelfth century. Then the circular arch was replaced by the pointed arch, and there arose what is known as the First Pointed Style. This style shows considerable advance in the vigour and treatment of the ornamentation, the mouldings and foliage begun by the Normans being now greatly improved. The windows were invariably pointed, narrow and lofty, giving an effect of great spaciousness with height. The nave in Dunblane Cathedral is a good example of this style.

The Middle Pointed or Decorative Style prevailed in Scotland from the middle of the fourteenth to the middle of the fifteenth century. The details now became much lighter and more ornate. The windows were enlarged, and in the tracery the eye was led to dwell more on the outlines of the bars than on the form of the aperture as in the earlier style. Parts of Dunkeld Cathedral show good examples of this style.

In the Third or Late Pointed Style the geometric tracery of the earlier periods has assumed a very flowing character. The tracery was called flamboyant because of the flame-like shape of the bars. This feature is more characteristic of the French architecture of the period than of the English. In England the tracery assumed a rigid form and the mullions of the windows were carried up in straight lines from the sill to the arch, so that the style was called Perpendicular. The exterior of the Scottish churches of this period is marked by heavy buttresses often with a great many set-offs. The semi-circular arch of the earlier styles is also frequently used in doors, pier arches, and clerestories, as in Dunkeld Cathedral.   In Scotland the buildings of this period consist largely of collegiate edifices.

The precise age of Dunblane Cathedral is not known but it is believed to have been founded by David I towards the end of his reign. The entire building, with the exception of the tower, is in the Early Pointed style of about the thirteenth century. The four lower stages of the tower, which stands on the south aisle of the nave, are all that remain of the original Norman structure. It has a shafted doorway and rib-vaulted basement story. The nave is almost entirely pure First Pointed. The west front of the Cathedral is particularly fine. Over the doorway are three very narrow, two-light windows, with quatrefoils at the heads of the two side windows and a cinquefoil at the head of the central one. Above these is a vesica set with a fringe of bay leaves. In speaking to an Edinburgh audience of this portion of the building, Ruskin said, " Do you recollect the west window of your own Dunblane Abbey ? It is acknow-ledged to be beautiful by the most careless observer. And

Dunblane Cathedral

why beautiful ? Simply because in the great contours it has the form of a forest leaf, and because in its decoration it has nothing but forest leaves. He was no common man who designed that Cathedral of Dunblane. I know nothing so perfect in its simplicity and so beautiful so far as it reaches, in all the Gothic with which I am acquainted. And just in proportion to his power of mind, that man was content to work under Nature's teaching, and instead of putting a merely formal dog-tooth, as everybody else did at the time, he went down to the woody bank of the sweet river beneath the rocks on which he was building and he took up a few of the fallen leaves that lay by it, and he set them in his arch side by side for ever."

Dunkeld Cathedral comprises a seven-bayed nave, a four-bayed aisle-less choir, a rectangular chapter-house and a massive tower. All the parts are of Second Pointed style with the exception of the choir, which exhibits some portions of First Pointed work. The nave shows many features of the French Flamboyant, especially the great west window, which judging from the remaining fragments of its tracery must have been of a particularly florid design. Dunkeld appears first "as a Culdee church, founded shortly before the accession of the Scottish kings to the Pictish throne." Here Kenneth MacAlpin transferred the relics of St Columba and built a church to be the mother-church of Celtic Christianity. The abbot of Dunkeld was also bishop of Fortrenn. When the bishopric was transferred to Abernethy, the abbot of Dunkeld came to be a layman. Early in the twelfth century Alexander I established a Roman bishopric at Dunkeld. Sometime about 1320 the present building was commenced and was finished about 1500. After the Reformation the choir was transformed into the parish church. One of the most exciting episodes in its history was its defence in 1689 by a small band of Cameronians under Clelland against 5000 Highlanders.

The Church of St John, Perth—originally the Kirk

Dunkeld Cathedral

of the Holy Cross of St John the Baptist—was in the twelfth century one of the most magnificent churches in Scotland.   As it now stands, it is of various dates, the western part being the oldest.   It is cruciform with a square central tower surmounted by an oak spire covered with lead.   In 1227 the church was granted to the

St John's Church, Perth

monks of Dunfermline, who allowed it to fall into disrepair.   Bruce ordered its restoration in 1328, but that ceased with his death.   In the fifteenth century the magistrates completely renovated the eastern portion. The church remained fairly complete till 1559, when the "rascal multitude" wrought great destruction on it.

Throughout Perthshire there are the remains of numerous ecclesiastical edifices which in their day must have been structures of great importance. There were Abbeys, for example, at Scone, Coupar-Angus and Inchaffray, and Collegiate Churches at Methven and Tullibardine, but little more than fragments of these can now be seen.

Prior to the Reformation there were in Perth and the neighbourhood numerous important monasteries and other religious houses of which no trace has now been left. The Dominican or Blackfriars Monastery, situated on the north side of the town, was founded by Alexander II in 1231. The Scottish kings frequently took up their abode in it, for which reason it was sometimes spoken of as a palace. There was a church in connection with the monastery, in which several parliaments were held. The Carmelite or Whitefriars Monastery at Tulilum goes back to the reign of Alexander III. The Charter-house or Carthusian Monastery, the only house of its order in Scotland, was situated near the spot where James VI's Hospital now stands, and owed its origin to James I and Jane his Queen in 1429. The Franciscan or Greyfriars Monastery, which stood on the present site of Greyfriars Churchyard, was founded by Lord Oliphant in 1460.

18.   Architecture—(b) Castellated.

The mansions of the Scottish nobility were, till comparatively recent times, mostly feudal strongholds; and numerous fine examples of these are to be found within the boundaries of the county.   Some of them are still in a good state of preservation while others are now in ruins. These castles tell of the habits of a people who, inured to war, had little care for their ordinary dwellings so long as their cattle and movable possessions could be safely placed beyond the ravages of the predatory invader.   The history of the county shows how the invader could hope to meet with little plunder until he had reduced such places of strength, behind which the natives were entrenched and from which they continually issued to harass their foe.

On the summit of Dunsinane Hill there are vestiges of a strong ancient fort, which according to Shakespeare and tradition is the Castle of Macbeth.   In 1857, while excavations were being made on the site, a doorway and underground chamber were discovered.

Huntingtower Castle is situated on the Crieff Road about 2½ miles from Perth.   Originally called Ruthven Castle, it belonged to the Earls of Gowrie.   Historically it is interesting as being the scene of the incident known as the Raid of Ruthven (1582) in which the Earl of Gowrie played a prominent part.   The castle consists of two massive square towers separated by a space of nine feet called the "Maiden's Leap."   The story, according to Pennant, was that the first earl's daughter leapt it one night when her mother had all but surprised her with her lover, with whom she eloped next morning.

Doune Castle stands on a steep, woody, greensward peninsula at the junction of the Ardoch Burn with the river Teith.   Though now roofless and ruinous, it is

Doune Castle

still a majestic pile, with its two massive square towers, turrets and high embattled walls.   The interior is full of long winding stairs, corridors, passages and deep gloomy vaults, which are well worthy of a careful examination. The castle is said to have been built about 500 years ago by Murdoch, Duke of Albany.   It was to this castle that the hero of *Waverley* was borne by his Highland captors.

Elcho Castle is a fine ruin, which some time ago was re-roofed, so that it may yet outlive many generations of mankind.   It is situated on the right bank of the Tay about four miles below Perth.   There is no inscription

Elcho Castle

upon the present castle to tell when it was erected though it must be of considerable antiquity.   The style of its architecture seems, however, to show that a still older and equally strong structure stood upon the same ground, but wanting the decorative details of cornices, architraves and abutments, which enrich the present building. Elcho Castle makes its first appearance in history, when Wallace and his heroic band lodged here in November 1296, previous to his attempt to recover Perth from the English.

Drummond Castle, near Crieff, is the Scottish seat of

Drummond Castle

the Earl of Ancaster. On the castle rock stand two structures of widely different periods. The ancient castle was built in 1491 by John, first Lord Drummond. It was often visited by James IV, and twice by Queen Mary in 1566. During the campaign of Cromwell it was almost demolished by his troops, and fell into even greater dilapidation after the Revolution of 1688. The remains were greatly strengthened and garrisoned in 1715 by the royal troops. Jane Gordon, Duchess of Perth, who was an ardent supporter of the House of Stuart, caused the walls to be levelled to the foundations during the "Forty-five" lest it should again fall into the hands of the royal troops. The castle was partly rebuilt in 1842 and a portion is used as an armoury containing a large collection of Celtic claymores, battle-axes and targets.

Kinclaven Castle crowns a strong and picturesque eminence upon the right bank of the Tay opposite the point where the Isla flows into it. It it said to have been built by Malcolm Canmore in the eleventh century, and for a long period of years it was a royal residence. Wallace won it from the English in 1296 or 1297, when, according to Henry the Minstrel, it was commanded by Sir James Butler, "ane agit cruell knicht." Visiting Perth under disguise, Wallace learned that the garrison was to be strengthened by 90 horsemen from Perth. He hastened to Kinclaven and attacked the castle with a handful of men, putting the entire garrison to the sword. Henry describes the engagement, and the flight of the English towards the castle, where

> "Few men of fenss was left that place to kepe,
> Wemen and preistis upon the wall can wepe:
> For weill thai wend the flearis was thar lord;
> To tak him in thai maid thaim redy ford,
> Leit doun the bryg, kest wp the yettis wide.
> The frayit folk entrit, and durst nocht byde."

The castle, now in ruins, must have been abandoned for many centuries as old fruit trees are growing in the courtyard.

Castle Huntly near Inchture Station in the Carse of Gowrie occupies a conspicuous position on a precipitous rock that rises, on all sides except the north-west, sheer from the dead level of the Carse. The castle was built

Castle Huntly

about the beginning of the fifteenth century. About the end of the eighteenth century it was converted into a modern residence with wings, battlements, round towers and corner turrets. The stone of which the castle is built was obtained from the great quarry of Kingoodie, which, by the way, also supplied the blocks for the Waterloo Bridge over the Thames. The interior combines all the features of a modern residence and an ancient stronghold, the rock-dungeon being particularly gruesome. The castle has been described as one of the best specimens of an old baronial residence in Scotland and as one of the most remarkable combinations of old and modern masonry

Tower of Kinnaird, Carse of Gowrie

in the kingdom. It was built by the second Baron Gray and tradition says that he named it after a daughter of the Earl of Huntly. Afterwards purchased by the Earl of Strathmore, it was known as Castle Lyon and subsequently came into the possession of the Paterson family. In the Carse of Gowrie also stands the Tower of Kinnaird, a square building of freestone which was visited by James VI in 1617.

The Castle of Inchbrakie near Abercairny is an ancient ruin surrounded by a moat. It is said to have been destroyed by Oliver Cromwell to punish the proprietor, Patrick Graeme, for his adherence to the Royalist Cause. Another old castle in the Crieff neighbourhood is Innerpeffray, beautifully situated on the banks of the Earn. Built in the sixteenth century, it has offered a stout resistance to the ravages of time. Its walls with a staircase and some of its apartments are still in a fair state of preservation.

The castles in the Highland area are neither so numerous nor so important as in the Lowland. Grandtully Castle, which is situated about three miles to the north-east of Aberfeldy, is a fine example of the Scottish baronial, dating from 1560. It has recently been restored in the old style. The main building consists of two five-storied towers, whose walls are nine feet in thickness. Later additions of gables and pepper-box turrets have been made.

Garth Castle stands on a bold promontory formed by two branches of the Keltney Burn about 2½ miles north-east of Fortingall. The keep or tower, of which only three sides remain, is from 60 to 70 feet high measured from the ground inside. The staircase, which gives entrance to the various stories, occupies the centre of one of the walls, which vary from six to seven feet in thickness. The position of the staircase seems to indicate that the castle must be of considerable antiquity as no such arrangement is to be met with in castles of a comparatively recent period. The castle was completely restored in accordance with the original plan by the late Sir Donald Currie, when it fell into his possession. During the latter half of the fourteenth century the castle was a stronghold of Alexander Stewart, Earl of Buchan, the "Wolf of Badenoch."

Meggernie Castle stands on the left bank of the river Lyon at the head of the inhabited portion of Glen Lyon. This picturesque castle, built in the simple and severe baronial style, is in keeping with its mountainous surroundings. The older portion is a large square tower of the fifteenth century, with a high peaked roof and four corner bartizans. The interior contains dungeons, secret apartments, and strongly barred doors as well as a number of relics, all of which are characteristic of the period when it was built. Other castles in the Highlands are Castle Dubh—in ruins—near Moulin village, supposed to have been built in the eleventh or the twelfth century; Castle McNiel, an old feudal tower near Cashlie in Glen Lyon; and Finlarig Castle, Killin, the ancestral seat of the Campbells of Lochow, from whom the family of Breadalbane takes its origin. The interior of this castle shows the dungeons with the old fetters still fastened to the walls.

19.   Architecture — (c) **Municipal and Domestic.**

The Municipal Buildings of Perth, at the north corner of High Street and Tay Street, form a fine edifice in the Tudor style, and include a copy of the old tower of

St Mary, a prominent feature of the town-hall and police station which formerly stood on the same spot. The spacious council chamber contains five beautiful stained-glass windows, the subjects of which comprise scenes from

Fair Maid's House, Perth

Sir Walter Scott's *Fair Maid of Perth*, the capture of Perth by King Robert the Bruce in 1311, and representations of Queen Victoria and Prince Consort. Further south and also facing the Tay stand the County Buildings, erected in 1819 on the site of Gowrie House. Designed after the style of the Parthenon at Athens, they are considered by competent judges to be a model of good taste—correct, simple and dignified, yet not deficient in ornament. The handsome buildings in the Scotch baronial style running along Canal Street and Tay Street include the Opera House and Public Halls, the Natural History Museum and other institutions.   One of the oldest

Scone Palace

houses in the city is that known as the " Fair ¦Maid of Perth's " House, at the corner of Blackfriars Wynd and Curfew Row. Here Simon Glover, the father of the " Fair Maid," is supposed to have resided. Formerly a niche in a corner of the house held an image of St Bartholomew, the patron saint of the glovers' incorporation, who at one time met in this house.

Only a few of the mansions and private seats in the county can be described here. Scone Palace, which belongs to the Earl of Mansfield, was built, in 1803, on the site of the old palace. Facing the river Tay and surrounded by beautiful gardens and woods, it is a castellated edifice of imposing dimensions situated in a park extending to upwards of a thousand acres. On the south front is Queen Mary's tree, said to have been planted by her own hands, while near the river there is a magnificent oak planted by James VI. The interior of the palace can boast of priceless treasures of painting and sculpture as well as historical relics. The furniture includes a bed which belonged to James VI, and another the hangings of which were worked by Queen Mary when imprisoned in Loch Leven Castle. The music hall occupies the site of the old great hall where the coronation of the Scottish kings took place.

Delightfully situated in an undulating woodland at the base of Kinnoull Hill, is the Castle of Kinfauns, a seat of the Earl of Moray. It is a vast modern castellated building with a central tower 84 feet high and a noble portico at the entrance. One of the relics of the olden times preserved in the castle is the two-handed sword of Thomas de Longueville, the compatriot of Wallace. The sword is a formidable weapon, measuring five feet nine inches long and two feet six inches broad at the hilt.

Another fine example of domestic architecture is Dupplin Castle, the seat of Sir John Dewar, near Forteviot Station, which is the successor of an older castle, destroyed by fire in 1827. A splendid Tudor structure, it commands a magnificent view of nearly the whole of Strathearn. The collection of books in the library is famous for many rare editions of the classics.

Rossie Priory, finely situated on Rossie Hill, looks down upon the Carse of Gowrie and surveys a wide scene of singular beauty. Built a century ago, it is the seat

Rossie Priory

of Lord Kinnaird. It is an imposing pile of monastic appearance, spacious and elegant, and contains a valuable collection of Roman antiquities.

In Highland Perthshire there are many fine houses belonging to the nobles and gentry, most amid very picturesque scenery. Taymouth Castle, the seat of the Marquis of Breadalbane, is at Kenmore near the exit of the Tay from Loch Tay. It is built of a light grey stone (chlorite schist), very soft and easily dressed when taken from the quarry—so soft that it can be cut with a knife or axe—yet remarkably hard after being some time exposed, and very durable. The present castle, built where stood the Castle of Balloch, consists of four stories with round towers at the angles and a massive quadrangular tower

Taymouth Castle

rising in the centre of the main building to a height of 150 feet. The Queen's room, the banner hall and the Chinese room are gorgeously fitted up. The castle contains paintings by Titian, Rubens, Salvator Rosa, Carracci, Teniers, Vandyke and other great masters. The magnificent library contains many rare and valuable works. The pleasure grounds comprise a circuit of fully 13 miles.

Blair Castle, near the mouth of Glen Tilt, about three-quarters of a mile north-west of the village of Blair Atholl, is a fine four-storied mansion, turreted and battlemented in the Scotch baronial style.   The present edifice has gradually grown up round the original part called Cumins Tower, built by John de Strathbogie, grandson of Macduff, the sixth Earl of Fife, in the thirteenth century.   The

Blair Castle

castle has many historical associations.   It is supposed that James V, in 1529, and Mary Queen of Scots, in 1564, must have visited it when hunting in Glen Tilt. It was occupied by Montrose in 1644; and in 1653 it was taken by one of Cromwell's officers and destroyed. In 1689 it was garrisoned by Dundee previous to the Battle of Killiecrankie.   The young Pretender lodged in the castle for three nights during the month of August, 1745. In March of the following year it was held for a fortnight by Sir Andrew Agnew for the government against Lord George Murray, the Duke of Atholl's brother. The garrison was reduced to great straits but was saved by the withdrawal of Lord George under orders from headquarters.

It will readily be understood that the architecture of the county has been affected to a considerable extent by the nature of the materials available for building. In the valley of Strathmore good building stone can usually be obtained from different parts of the Old Red Sandstone formation. Thus along the Highland boundary such towns and villages as Blairgowrie, Comrie, Crieff and Aberfoyle have largely availed themselves of the finer and more suitable beds of conglomerate that are so extensively developed in the lower parts of the Old Red Sandstone. Along the central and southern districts of Strathmore fine-grained white and red sandstones occur in the higher parts of the Old Red Sandstone, and these have been extensively used in the building of the city of Perth, and of Coupar-Angus, Auchterarder and Dunblane. Many of the buildings in this area have been roofed with slates from the Birnam, Craiglea, or Aberfoyle quarries.

In the Highland area, no sandstone being available, the inhabitants have had to make use of the most suitable materials in their neighbourhood ; and as the Highland schists cannot be worked with the same ease into decorative designs as the sandstones, little or no ornament is found in the Highland buildings. Even to hew out simple blocks of schist is a much more difficult and laborious task than the hewing of sandstone. Notwithstanding this, many large and substantial edifices have been erected out of such bands as the Ben Ledi Grits, the Green Beds and the Moine Schists. The town of Aberfeldy has been largely built out of the Green Beds of the neighbourhood.

Cottages at Killin

This stone, of a dark-green or greenish-grey colour, presents a very handsome appearance and can be fairly easily dressed. Another stone which has been used in the Aberfeldy district is the talcose schist of Bolfracks Quarry. This stone is homogeneous in structure, dark green in colour, soft and somewhat soapy to the touch. Easily wrought, it can be used for rough carving while

it offers a greater resistance to the weather than any of
the other stones in the district.   The ornamental work
on the church tower at Kenmore and the pillars upon
the Tay Bridge at Aberfeldy have been executed in this
stone.   In the more north-westerly parts of Highland
Perthshire, as at Blair Atholl and Kinloch Rannoch,
certain siliceous bands in the Moine rocks are chiefly
used for building.   In some places handsome villas and
other edifices have been built of dolerite from the long
east and west dykes.   The black graphitic schist of Ben
Lawers has also been used for roofing.

20. Communications—Past and Present.

Throughout Perthshire there are numerous evidences
of the former existence of Roman roads.   Even before
Roman days, the original inhabitants of the county no
doubt had tracks or paths of which no traces now remain.
Stretches of Roman roads have been identified at Gask,
Abernethy, Meigle, Cargill and other localities.   At Gask
the Roman road from Ardoch to Orrea, which, says
W. F. Skene, lay near the junction of the Earn with the
Tay, intersects the parish running along the ridge which
forms the highest ground.   The road, about 20 feet broad,
is formed of a causeway of rough stones laid closely
together; and along the side of the road were stations,
remains of which are still visible.   At Meigle traces have
been found of the great Roman road leading from Coupar
to Battle Dykes; it passed near the camp at Cardean.

Speaking of the road from Ardoch, Dr James Browne
says, "Having crossed the Tay by means of the wooden

General Wade's Road, Glen Ogle

(*With other roads and railway line*)

bridge (about two miles north of Perth), the Roman road
proceeded up the east side of the river, and passed through

the centre of the camp at Grassy Walls. From this position the remains of the road are distinctly visible for a mile
up to Gallyhead, on the west of which it passed and went
on by Invertrust to Nether Collin, where it again becomes
apparent and continues distinct to the eye for two miles
and a half, passing on to Drichmuir and Byres. From
thence the road stretched forward in a north-east direction,
passing between Blairhead and Gilwell to Woodhead, and
thence pushing on by Newbigging and Gallowhill on the
right, it descends Leyston-moor, and passing that village,
it proceeds forward to the Roman camp at Coupar-Angus
about eleven and a half miles from Orrea."

Soon after the Rebellion of 1715, General Wade was
sent to the Highlands to make an inquiry into the condition of the country and its people. Shortly afterwards
he began to make a system of metalled roads, which can
be seen in different parts of the county. In Glen Ogle,
as elsewhere, Wade's road has fallen into disuse, and its
bridges have been allowed to crumble to pieces, yet it
can be distinctly traced intersecting the present road in its
course through the glen.

His roads greatly improved the means of communica-
tion in the Highlands and his work has been commended
in the well-known lines:

" Had you seen but these roads before they were made
  You would have held up your hands and blessed General
    Wade."

The present roads of the county are well constructed
and well kept, together with the bridges over which they

pass. The Edinburgh road, which passes through Queens-ferry, Dunfermline and Kinross, enters the county a few miles to the south-east of Perth, which it reaches across the South Inch. The Glasgow road passes through Stirling, Dunblane and Auchterarder. The road from Dundee approaches from the east passing through the Carse of Gowrie. A fourth road from Comrie, Crieff and Methven enters the town on the north-west. The great Highland road starting at Perth runs along the valley of the Tay, the Tummel and the Garry, passing through Bankfoot, Dunkeld, Pitlochry, the Pass of Killie-crankie and Blair Atholl. This important road with its numerous side roads, parliamentary roads and bridges was planned and carried out by Telford in the nineteenth century. From Perth a road runs to Forfar and Aber-deen through Strathmore by way of Coupar-Angus. At Coupar-Angus a branch forks off to the north leading through Blairgowrie, Glen Shee and Glen Beg into Braemar. In addition to these main thoroughfares there are numerous smaller roads by which the surrounding districts are made accessible.

The principal railway systems in the shire are the Caledonian, the North British, the Highland, the Callan-der and Oban, and the West Highland. The Caledonian main line enters the county a little to the south of the Bridge of Allan and runs north-eastwards through Dun-blane, Auchterarder, Perth, Stanley to Coupar-Angus, a distance of 46 miles. A branch line runs off from Crieff Junction to Crieff, Comrie and St Fillans, joining the Callander and Oban line at Balquhidder. From

Perth a branch line runs to Crieff through Methven, while another runs to Dundee through the Carse of Gowrie. Other branch lines run from Coupar-Angus to Blairgowrie and from Meigle to Alyth. From Dunblane a branch is thrown off to Doune and Callander, where it joins the Callander and Oban line, which continues up Strathyre, Glen Ogle and Glen Dochart to Tyndrum. The North British line enters the shire near Newburgh in Fife, and proceeds to Perth by the Bridge of Earn, a distance of nine miles. A direct route between Perth and Edinburgh by the Forth Bridge passes through Glen Farg. The same company's Aberfoyle branch of the Forth and Clyde line lies for the most part in Perthshire. The West Highland Railway from Craigendoran, after passing up Glen Falloch and Strath Fillan, leaves the shire for a short distance but re-enters it near the headwaters of the river Leven. The Highland Railway branches off from the Caledonian at Stanley and ascends the valleys of the Tay, the Tummel and the Garry, with stations at Dunkeld, Ballinluig, Pitlochry, Pass of Killiecrankie, Blair Atholl and Dalnaspidal. It is 51 miles in length from Perth to Dalnaspidal. A branch eight miles in length runs from Ballinluig to Aberfeldy.

Steamboats ply on Loch Tay from Killin at the head of the loch to Kenmore at the foot. The piers on the north side of the loch are Lawers and Fernan and on the south side Ardeonaig and Ardtalnaig. They have connections with the Callander and Oban Railway, and with a coach that runs between Kenmore and Aberfeldy.

Steamboats also ply on Loch Katrine between Stronach-lachar and the Trossachs, and on the Tay between Perth and Dundee. Some of the larger Highland lochs have ferries at various points.

There used to be a number of ferries on the Tay between Perth and Dunkeld, but these are gradually being replaced by bridges. The old ferry boats were

Kinclaven Ferry

worked by chains; and besides passengers, they were able to carry light vehicles. Perhaps one of the best known of these ferries was that at Kinclaven, but it has been superseded by a handsome stone bridge. Along the Firth of Tay between Perth and Dundee there were many little harbours which have now fallen into disuse. These were situated at Kingoodie, which was

formed for the exportation of sandstone quarried in the
neighbourhood, Powgavie south of Inchture, and Port
Allen near Errol. These ancient harbours, now silted
up and overgrown with weeds, are witnesses of bygone
days when the shortest route for the transference of
merchandise from Fife to Strathmore lay across the Tay
and from thence, by the numerous roads through the
Sidlaws, to Blairgowrie, Coupar-Angus, Meigle, Alyth,
Kirriemuir and Forfar.

21. Administration and Divisions— Ancient and Modern.

In early times the county was divided into districts
under the jurisdiction of hereditary governors, viz.
Menteith, Breadalbane, Strathearn, Methven, Atholl,
Strathardle, Glenshee, Stormont, Gowrie and Perth.
Though these districts have no longer any judicial or
civil existence, yet the names are in constant use in
referring to the geography of the shire. Menteith and
Strathearn were stewartries, Breadalbane a bailiary with
separate jurisdiction of its earls, Methven a separate
regality, and Atholl a regality of very great extent. The
judicial president of the county was the shire-reeve or
sheriff, who was the official deputy of the crown and was
responsible for the enforcement of law and order. This
office was hereditary and was usually in the hands of some
leading landowner, not necessarily possessed of any legal
qualifications. By degrees these hereditary powers were

reduced, till after the "Forty-five" the Act of 1747 was passed entirely abolishing them, and appointments to the office of sheriff were made on the present method.

Three distinct classes of burghs have existed in Scotland from very early times—Royal Burghs, Burghs of Regality, and Burghs of Barony. The Royal Burgh is the most complete and perfect form of burghal constitution in Scotland. This corporate body, created by a charter from the crown, has the right of self-government by a magistracy and council, and for the payment of fee, farm rent or burgh mail possesses many important privileges. The charter of Perth (the only royal burgh in the shire) is dated 12 October 1210, in the reign of William the Lion, who, in a subsequent charter, is styled "The founder and instaurator of our said royal burgh of Perth after the visitation and ruin thereof by the inundation of the said flood and river Tay."

The Burgh of Barony is another kind of municipal corporation, of which Alyth, Crieff and Blairgowrie are examples. A Burgh of Barony consists of the inhabitants of a definite tract of land within the barony, placed under the authority of magistrates and a council, whose election is vested either in the baron superior, or is in the hands of the inhabitants themselves, according to the terms of the charter. A later kind of municipal government is a Police Burgh, which is constituted by the sheriff for the purposes of improvement and police, and is granted to populous places, with boundaries fixed in terms of statute, the local authority being the police commissioners.

The county of Perth is governed by a lord-lieutenant, a vice-lieutenant, and a large number of deputy-lieutenants and justices of the peace. The law is administered by a sheriff and sheriff-substitute.

The most important administrative body in the shire is the County Council, which looks after the finances, roads, bridges, public health, and the general administration. By the Redistribution of Seats Act (1885) Perthshire was divided into two divisions—the Eastern and the Western, each of which returns one member to Parliament.

The county was divided at an early period for ecclesiastical purposes into parishes, of which there are now 71, as follows—Aberdalgie, Aberfoyle, Abernethy, Abernyte, Alyth, Ardoch, Arngask, Auchterarder, Auchtergaven, Balquhidder, Bendochy, Blackford, Blair Atholl, Blairgowrie, Callander, Caputh, Cargill, Clunie, Collace, Comrie, Coupar-Angus, Crieff, Dron, Dull, Dumbarney, Dunblane, Dunkeld and Dowally, Dunning, Errol, Findo-Gask, Forgandenny, Forteviot, Fortingall, Fowlis-Wester, Glendevon, Inchture, Kenmore, Killin, Kilmadoch, Kilspindie, Kincardine, Kinclaven, Kinfauns, Kinnaird, Kinnoull, Kirkmichael, Lecropt, Lethendy and Kinloch, Little Dunkeld, Logierait, Longforgan, Madderty, Meigle, Methven, Moneydie, Monzie, Monzievaird and Strowan, Moulin, Muckart, Muthill, Perth, Port of Menteith, Rattray, Redgorton, Rhynd, St Madoes, St Martins, Scone, Tibbermore, Trinity Gask, Weem. These parishes are divided among the presbyteries of Dunkeld, Weem, Perth, Auchterarder and Dunblane in the synod of Perth and

Stirling; the presbyteries of Meigle and Dundee in the synod of Angus and Mearns; and the presbytery of Kinross in the synod of Fife. In many cases parishes are found in detached portions as Dull, Weem, Kenmore, Killin, Caputh, Kinnoull, etc.

A number of the parishes are of great extent, as Blair Atholl, which is 30 miles by 18, and Fortingall which is 37 miles by 17. It is interesting to compare the size of these two parishes with the county of Clackmannan, the smallest in Scotland, which is only nine and three-quarter miles in greatest length and eight and three-quarter in greatest breadth.

Within the Highland area there are lands common to several parishes, such as Blair Atholl, Logierait, and Fortingall. There are also some 2326 acres in the county not claimed by any parish.

Formerly Perthshire was divided into the bishoprics of Dunkeld, Perth and Dunblane, and accordingly these towns have been designated cities. The division of the county as given above into parishes is used for such civil purposes as the administration of the poor law, the registration of births, deaths and marriages, and taxation, as well as for the parochial and the political franchise.

Since the great Education Act of 1872, the management of education in Scotland has been entrusted to School Boards, which are elected every three years by the ratepayers. These boards are established in every parish and burgh in the country, and are responsible for the management of the schools and the appointment of teachers. Compulsory attendance for all children between the ages of five and fourteen years is universal in Scotland. This primary education is free.

Secondary Education is supplied by endowed schools, Higher Grade Public Schools, and by proprietary and other schools. Trinity College, Glenalmond, now one of

Glenalmond School

the leading schools of Scotland, was founded in 1847. In 1888, the Education Department instituted a system of examinations for leaving certificates, which has been taken advantage of by all the best secondary schools in the county.

22. The Roll of Honour.

It is curious to note that the Perthshire roll of honour is comparatively small. Though its scenery has inspired such great poets as Scott, Wordsworth and Burns, yet the shire has not produced any poet of the first order. Though it has reared the rank and file of such famous regiments as the "Black Watch," yet it has given us no outstanding warriors but Lord Lynedoch. Though it has supplied the material for the deductions of such pioneers in geological science as Hutton and Playfair, yet it has been the birthplace of no great geologist with the exception of Dr Croll. We cannot here enter into a discussion of the various factors that have governed the distribution of genius in our country, but it must be frankly admitted that Perthshire shows a remarkable dearth of prominent men.

Among the Perthshire poets the following may be mentioned. Henry Adamson, born in Perth, was author of *The Muses Threnodie with a description of Perth and an account of the Gowrie Conspiracy*. Robert Nicol, poet and journalist, who has been described as Scotland's second Burns, was born in Auchtergaven parish in 1814, and published *Poems and Lyrics* in 1833. Perhaps the most distinguished of the Perthshire poets, and certainly the most widely known, was the Baroness Caroline Oliphant Nairne, song and ballad writer. She was the daughter of Laurence Oliphant, and was born in the old mansion house of Gask in 1766. Towards the close of the

Lady Nairne

eighteenth century her songs, many of which were
Jacobite, were sung in every district of the kingdom.
They include *Will ye no come back again?*, *The Laird o'
Cockpen*, *The Land o' the Leal*, *The Auld Hoose*, and
Caller Herrin'. David Malloch, born at Crieff about
the year 1700, was another of the Perthshire minor
poets. Settling in England, where he changed his name
to Mallet, he wrote tragedies, as *Elvira*, and the ballad
William and Margaret. On weak grounds, the author-
ship of *Rule Britannia* has been claimed for him.

Duncan Ban Macintyre, or the fair-haired Gaelic
poet, was born in Argyllshire in 1724, but much of his
poetry refers to Perthshire. In his early life he was
employed as a forester on the Breadalbane forest of the
Blackmount, and his poem on *Beinn Doireann*, in that
district, is considered to be one of the finest examples
of modern Gaelic poetry. One of Duncan's best pieces,
The Last Farewell to the Hills, was written when he was
seventy-eight.

Another Gaelic poet was Dugald Buchanan, born in
Balquhidder. At Kinloch-Rannoch he settled as school-
master and catechist, and there he wrote the most of his
hymns and poems.

Painting is represented by Thomas Duncan, born at
Kinclaven in 1807. Perhaps his best known work is
"Prince Charles entering Edinburgh." Other of his
pictures are "Martyrdom of John Brown of Priesthill,"
"Abdication of Queen Mary," and "Wishart dispensing
the Sacrament."

The only sculptor of note born in the county was

Lawrence Macdonald, born at Gask in 1798. He produced many fine busts, and well-known classic groups, as "Ajax bearing the dead Patroclus," "Ulysses recognised by his Dog," and others.

The only eminent musician is Neil Gow, born at Inver near Dunkeld in 1727. He was renowned as a violinist and composer, and in his day was considered to be unrivalled in the playing of strathspeys and reels.

Among the literary men of the county notice must be taken of George Gilfillan, author and clergyman, who was born at Comrie in 1813. Two of his principal works are *The Bards of the Bible* and *The Martyrs of the Scottish Covenants.*

In the domain of science Perthshire has few outstanding names. The distinguished physicist and geologist, Dr Croll, was born at Little Whitefield in 1821. His life work consisted principally in an endeavour to find a true cause for the great extension of snow and ice in northern Europe during the Ice Age, for which purpose he invoked the aid of astronomical and terrestrial physics. His theory was received with much enthusiasm by geologists. His chief works are *Climate and Time in their Geological Relations* and *Discussions in Climate and Cosmology.* David Douglass, botanist and traveller, was born at Scone in 1799. He assisted Dr Hooker in collecting the materials for his *Flora Scotica.* In 1823 he was sent to the United States on a botanical expedition by the London Horticultural Society. He also surveyed the Columbia River District, 1824–30.

The only soldier of note that the county can claim as

Neil Gow

a son is General Sir Thomas Graham, Lord Lynedoch, born at Balgowan near Methven in the year 1748. He was aide-de-camp to Sir John Moore all through his

Dr James Croll

Peninsular campaign. Made Major-General in 1810, he took command of the British and Portuguese troops in Cadiz, then blockaded by the French. He afterwards joined Wellington's army, fought at Ciudad Rodrigo, at Badajoz, and at Vittoria, where he commanded the left wing.

General Sir David Baird was another eminent soldier closely connected with the shire, though not born in it. He took part in the storming of Seringapatam, and after serving in Egypt was made commander of an expedition to Cape Colony. He assisted Sir John Moore, whom he succeeded in command after Corunna.

William Murray, first Lord Mansfield, was born in Perth in 1705. A distinguished lawyer and statesman, he was successively Solicitor-General, Attorney-General, and Lord Chief Justice of the King's Bench.

Another distinguished legislator belonging to the county was the Hon. Alex Mackenzie, Premier of the Dominion of Canada, who was born in the village of Logierait in the year 1823. He was the son of a local mason, with no advantages of birth, fortune or education, but his ability and sterling character procured for him both fame and fortune. His five years' ministry, during which Lord Dufferin was Governor-General, has been described as the purest administration which Canada had experienced.

Associated with the county are the names of many eminent divines. Patrick Adamson, born at Perth in 1537, was made Archbishop of St Andrews in 1576. Donald Cargill, a Covenanting preacher, was born at Rattray about 1620, and ordained to the Barony Charge, Glasgow, 1655. He opposed Episcopacy and took to field-preaching. In 1681 he was seized at Covington, tried at Edinburgh, found guilty of treason, hanged and

Sir David Baird

beheaded. John Brown, born at Carpow near Abernethy, became Professor of Divinity under the Associate Synod. His *Self-Interpreting Bible* achieved considerable popularity. John Barclay, the founder of the Berean Sect, was born at Muthil in 1734.

Though the county has a comparatively small roll of honour, yet the number of distinguished names in literature and science which have been connected with it in some way or other is very considerable. Many of the scenes in Sir Walter Scott's *Lady of the Lake*, *Rob Roy* and *The Fair Maid of Perth* lie within the boundaries of Perthshire. Burns made a tour through Perthshire, and some of his most exquisite lyrics have been inspired by its scenery, as *The Birks o' Aberfeldy*, *The Humble Petition of Bruar Water* and *On Scaring some Waterfowl on Loch Turrit*. Many of the Jacobite songs are associated with the shire, chief among these being Hogg's *Cam ye by Atholl*. Wordsworth's *Stepping Westward* was suggested by an incident which occurred to him at Loch Katrine.

The scenery of Perthshire has been painted over and over again by many British and foreign artists, whom even to enumerate would be impossible. John Ruskin spent much of his childhood in Perth and he tells us his father's sister " lived at Bridgend and had a garden full of gooseberry bushes sloping down to the Tay, with a door opening to the water, which ran past it clear-brown over the pebbles three or four feet deep, an infinite thing for a child to look into."

The botany and the geology of Perthshire have also attracted many eminent scientific men to the county.

William, First Earl of Mansfield

The discovery and description of the granite veins in Glen Tilt by James Hutton, the founder of physical geology, form a most important event in the progress of geological science. It was on Schiehallion that Dr Maskelyne, Astronomer Royal, made those observations and experiments by which he ascertained the power of rock masses in attracting the pendulum and determined from the result the mean density of the earth. Maskelyne was followed by Professor Playfair and his calculations were so far corrected in a complete mineralogical survey of Schiehallion. Playfair was also a disciple of Hutton, and in his *Illustrations of the Huttonian Theory* explained and defended the great principles first advanced by his friend. Many of his illustrations are drawn from Perthshire and he must have made a careful examination of the rocks of the county.

In 1771 Pennant, naturalist and antiquary, published a *Tour in Scotland in* 1769, followed in 1774 by an account of another journey in Scotland. In them will be found a description of the topography and general conditions of the county of Perth at that time. M'Culloch, in his work on the *Highlands and Western Isles of Scotland* (1824), gives an account of Highland Perthshire, in which he deals principally with its scenic features.

23. THE CHIEF TOWNS AND VILLAGES OF PERTHSHIRE.

(The figures in brackets give the population in 1911, the asterisk
denoting parishes. The figures at the end of each section
are references to the pages in the text.)

Aberfeldy (1592), finely situated five miles from Taymouth
on Moness Burn, is a great tourist resort. At this point the Tay

Monument to Black Watch

is spanned by a five-arched bridge constructed by General Wade
in 1733. It was in a field to the south of the bridge that the
famous Black Watch regiment was first embodied in 1739.

According to Pennant there were within the area of Loch Tay and Glen Lyon at that time 1000 men capable of bearing arms— a striking contrast to the present population.   Aberfeldy is noted for the manufacture of tweeds, tartans, plaids, etc.   A short distance from the town are the celebrated falls of Moness, the scene of Burns's song *The Birks of Aberfeldy*.   At this point the Moness Burn makes a succession of leaps, falling about 100 feet

Aberfoyle

within a distance of a few hundred yards.   Some years ago a quarry was made in one of the basalt dykes above Gatehouse, and the stone, extensively used for mending roads, is conveyed by a ropeway to the railway station and deposited directly in the waggons.   (pp. 101, 124, 133, 138.)

Aberfoyle (1147)*, Gaelic *abhair-a-phuill*, "confluence of the pool," is the terminus of the Strathendrick Railway.   Situated in a region of glens, mountains, rivers, cascades and lakes, it lies on

the north bank of the Forth, here known as the Avondhu
(Black Water). It is closely associated with many of the scenes
in Sir Walter Scott's *Lady of the Lake* and *Rob Roy*. On rising
ground near the manse there are ten large stones in a circle with
a still larger one in the centre. These were originally upright
but have now fallen and been more or less buried in the ground.
From the stones the kirk of Aberfoyle was called the "Clachan."
Aberfoyle makes a convenient centre for visiting the Trossachs
and the numerous lochs in this region. About three miles to the
east of Aberfoyle is the Lake of Menteith. On the island of
Inchmahome are the ruins of a priory, which was the refuge of
Queen Mary as a child. See Dr John Brown's *Queen Mary's Child
Garden*. (pp. 16, 36, 64, 74, 88, 132.)

Abernethy (1267)[*] stands on the right bank of the Nethy,
eight miles from Perth. Perhaps the most remarkable feature
about the village is its fine round tower. It was once a Pictish
capital and a religious centre. (pp. 93, 103, 109, 134, 153.)

Aberuthven is a village situated in the north of Auchterarder parish. The ancient parish church dedicated to St Katlan
is now a roofless ruin. Near it stands the mausoleum of the
Dukes of Montrose.

Almondbank is a village lying about three quarters of
a mile north of a station of the same name on the Crieff and
Methven Railway. The inhabitants are principally employed in
the bleachfields of the neighbourhood.

Alyth (2937)[*] is a town situated on the Alyth Burn in the
east of Perthshire and on the confines of Forfarshire, in which
part of the town lies. On Barry Hill are the remains of a fort
which must have been of considerable strength and importance.
Alyth is a burgh of barony, under a charter of James III.
The parish church (1839) is of Norman structure with a lofty
spire, taking the place of the ancient Second Pointed church of

St Moloe. The houses are excellently built and the town has linen, flax, woollen and jute works, with bleaching, dyeing and calendering. (pp. 7, 16, 45, 138, 140, 141.)

Auchterarder (3175)*, Gaelic *uachdar-ard-thir*, "upper high land," so called because of its situation on the brow of a low hill, on the left bank of the Ruthven Water, 13¾ miles south-west of Perth. The town seems to have existed in the year 1200. It has the remains of a small castle, supposed to have been a

East Mill, Auchterarder

hunting seat of Malcolm Canmore (1088–93). It was a royal burgh and the chief burgh of Strathearn. About half a mile to the north are the old parish church and the well of St Mackerrok. After Sheriffmuir the Earl of Mar, fearing pursuit by the Duke of Argyll, burned Auchterarder to the ground. The town was closely connected with the events which led up to the Disruption of the Church of Scotland in 1843. It has cattle fairs, and manufactures of tartans and galas. (pp. 64, 82, 132, 137.)

Birnam

Bankfoot (2167)*, or Auchtergaven, situated on the high road between Perth and Dunkeld three and a quarter miles from Strathord station, is best known as the birthplace of the poet Robert Nicol, in whose memory a fine monument has been erected there.   The antiquities are St Bride's well, which marks the site of Logiebride Church, and a stone circle.   The staple industry is weaving, and many people are employed in the Airleywight linen works.   (pp. 16, 137.)

Birnam is a village, much frequented by tourists, with a station on the Highland Railway, 15½ miles from Perth.   Behind it rise the steep and rugged sides of Birnam Hill.   The royal forest immortalised in Shakespeare's *Macbeth* has long ago disappeared, and its place has been taken by a young and thriving plantation of firs and birches.   An oak and sycamore near the hotel are pointed out as the only survivors of the ancient forest. From the summit of the hill a magnificent panorama of Strathmore can be obtained.   (pp. 7, 16, 88, 132.)

Blackford (1374)*, on the northern base of the Ochils 17¾ miles south-west of Perth, near the confluence of the Danny Burn with Allan Water, is a clean and well-built village with several breweries as well as weaving and boot-and-shoe factories. When James IV returned from his coronation at Scone in 1488, as the treasurer's accounts state, 12 shillings was paid "quhen the king cum furth of Sanct Johnistone for a barrel of ayll at the Blackfurd."   (p. 103.)

Blair Atholl (1580)*, Gaelic "plain of the pleasant land," is a small village in the north of Perthshire at the confluence of the Garry and the Tilt.   At the mouth of Glen Tilt stands Blair Castle, the principal seat of the Duke of Atholl.   The district is famous for its wild natural beauty, for its great wealth of deer, grouse and salmon, for the general richness of its fauna and flora, and for its geological structure.   (pp. 38, 131, 134, 137, 138, 143.)

Blair Drummond is a small village on the right bank of the Teith six miles north-west of Stirling. The inhabitants are mostly employed on the Blair Drummond estate, which has some of the finest trees in the county. The neighbouring Kincardine Moss has yielded many interesting antiquarian remains, including a number of small Roman relics and two curious ancient wooden wheels. Blair Drummond Moss was

Blair Atholl

reclaimed by cutting canals through it in the direction of the river Forth. Water was raised to the canals from low levels by a species of undershot water wheel. The moss was cut and floated away to sea through the canals, at very little cost. The earth below was so rich that gravel had to be used to reduce it and lime to break it up. Afterwards the ground was colonised; and, according to John Ramsay, the crofters lived in great cheerfulness and content.

Blairgowrie (4319)*, Gaelic *blar-ghobhar*, "plain of the wild goats," on the river Ericht, is the terminus of a branch of the Caledonian Railway. During the last century it has risen from a mean collection of thatched houses to a well built residential town. Though situated at the opening of the passes that lead to Kirkmichael, Pitlochry and Braemar, it has but few historical associations of interest, unless with Skene we make Blair Hill the scene of the Battle of Mons Graupius. On the other side of the river is the tourist-haunted village of Rattray. At the Hatton of Rattray Donald Cargill, the martyr, was born. The scenery of the Ericht above Rattray is very picturesque, especially the deep cañon which the river has cut through the Old Red Sandstone conglomerates, upon a spur of which is built the mansion-house of Craighall. On an island in Loch Clunie, between Blairgowrie and Dunkeld, stand the ruins of Clunie Castle, a former residence of the Earl of Airlie. Blairgowrie has numerous thriving industries as flax-spinning, linen, carpet and jute-mills, sawmills, malt kilns, a farina factory and an agricultural implement factory. (pp. 7, 29, 81, 103, 104, 132, 137, 138, 140, 141.)

Bridge of Earn is a small village on the river Earn four miles south-east of Perth, named after the ancient bridge which here spanned the river. A mile to the south are the celebrated Pitcaithly mineral wells, supposed to be the oldest natural medicinal waters in the country. (pp. 40, 138.)

Callander (2215)*, on the Teith 16½ miles north-west of Stirling, lies in the centre of some of the finest hill and lake scenery, commanding fine views of Ben Ledi and the Pass of Leny. Callander is the great centre for tourists visiting the Trossachs and has also many summer residents. The village, which consists of one long street, is regularly built and contains numerous handsome villas. (pp. 7, 9, 16, 74, 88, 137, 138.)

Comrie (1745)*, finely situated on the left bank of the Earn at an altitude of 200 feet above sea-level, is a favourite summer and autumn resort. Comrie is celebrated chiefly on account of its earthquakes, which are regarded as due to the proximity of the great Highland boundary fault. On the summit of Dunmore Hill is a granite monument to the memory of Lord Melville, while another to Sir David Baird stands on a hillock a little to the east. (pp. 7, 16, 74, 132, 137, 148.)

Coupar-Angus (2749)*, a town of considerable antiquity, stands almost in the centre of Strathmore near the left bank of the Isla, about 15¾ miles from Perth. In the neighbourhood traces of the Romans are still to be seen, including what appears to have been an important camp. For a long time the town was in a somewhat decaying condition but within recent years it has made considerable progress and is now the centre of a flourishing trade. (pp. 78, 81, 116, 132, 136, 137, 138, 140.)

Crianlarich, a small village at the mouth of Strathfillan at an altitude of 522 feet above sea-level, has one station on the Callander and Oban, and one on the West Highland railway. In Strathfillan is the site of the ancient priory of St Fillan, founded by King Robert Bruce. The square-shaped bell of St Fillan, which lay on a gravestone till 1798, was stolen by an English traveller. It was afterwards restored and is now preserved in the Edinburgh Antiquarian Museum.

Crieff (6089)*, sometimes called the Capital of Upper Strathearn, stands on the left bank of the Earn. Through feudal times and till the abolition of hereditary jurisdiction, the town was the seat of the civil and criminal courts of the district. When Scott visited Crieff in 1796, the gallows still stood on the Gallow Hill, on the spot now marked by a lime tree. In the High Street stand a large sculptured stone and the old Town Cross. In 1716 the town was burned amid circumstances of great cruelty by the Chevalier's Highland adherents. Prince Charlie visited the town towards the close of the rebellion (1746), and it again narrowly escaped burning. The town has linen and woollen

Crieff

factories, leather works, barley, flour, bark, flax, linseed oil, saw and turning mills. Before 1720, when the Falkirk trysts were established, it had the largest cattle markets in Scotland. Its excellent climate attracts visitors and invalids. (pp. 7, 9, 74, 82, 103, 120, 124, 132, 137, 138, 141, 147.)

Deanston is a neat little hamlet on the right bank of the Teith about a mile west of Doune. It has extensive cotton-mills, founded in 1785. (p. 81.)

Doune, Gaelic "the hill," on the left bank of the Teith eight miles north-west of Stirling, consists of one main street and two smaller ones, which radiate from an old market cross. Doune was at one time famous for the manufacture of Highland pistols and sporrans. The pistol trade was introduced in 1646 by Thomas Cadell, and the weapons sold at from 4 to 24 guineas a pair. In 1745 Prince Charlie occupied Doune Castle. Twice a year cattle markets known as the Doune Fair are held. (pp. 101, 118, 138.)

Dunblane (4591)*, Gaelic "hill of Blane," an ancient city on the left bank of the Allan Water five miles north of Stirling and 28 south of Perth. The principal street is crooked and narrow and the houses mostly old. It is a favourite summer resort. Much of the interest of Dunblane centres round the remains of its ancient cathedral. The wool and worsted mills of Keir and Springbank give employment to a number of the inhabitants. (pp. 64, 82, 110, 111, 132, 137, 138, 143.)

Dunkeld (1081)*, Gaelic *dun chaillin*, "fort Caledonia" on the Tay 15 miles north of Perth, is an ancient city but now more like a village, entered by a handsome stone bridge built about the beginning of the last century by the Duke of Atholl. The view obtained from the bridge is very impressive. It would be hard indeed to conceive of anything more beautiful than the Tay flowing deep below amidst the noble oaks which skirt its banks and winding round the wooded pyramid of

Craig-y-Barns on the one side and under the wild acclivities of
Craig Vinean on the other, with the hoary cathedral nestling among

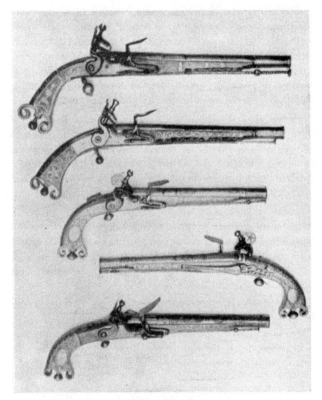

Doune Pistols

the trees upon a level stretch of haughland in front. The city
consists of two main streets with the old cathedral and Dunkeld
House at their heads. It has a large retail trade and is much

frequented by summer visitors. (pp. 23, 25, 29, 74, 88, 93, 105, 111, 113, 137, 138, 139, 143, 148.)

Dunning (1145)*, Gaelic *dunan*, "small fort," lies nine and a half miles to the south-west of Perth. Dunning was burned by the Highlanders in the retreat from Sheriffmuir. The parish church (early thirteenth century) is built in the Early English style of architecture. A good deal of the original building still remains, the massive square Norman tower being a striking object. The church was almost entirely rebuilt in 1810. (p. 101.)

Errol (2083)*, a small village on rising ground in the Carse of Gowrie eleven and a half miles east of Perth, is almost midway between Perth and Dundee and serves as a business centre for the Carse. The parish church, built in 1831, is a cruciform Norman structure with a massive square tower. (p. 140.)

Forteviot (549)*, seven miles south-west of Perth, was an ancient capital in Pictish times and later. The palace, of which no trace remains, stood on Halyhill to the north-west of the village. A sculptured stone which once stood here, having the figure of a king—supposed to be Alexander I—with crown and sceptre, and a bishop with mitre and crozier, is preserved in the Antiquarian Museum, Edinburgh. (pp. 93, 128.)

Gartmore is picturesquely situated on a hill between the river Forth and the Kelty Water. The old Peel of Gartfarren stood about a mile from Gartmore. In the neighbourhood is Flanders Moss, in which have been found embedded the remains of large trees, relics probably of the Great Caledonian Forest.

Huntingtower, three miles north-west of Perth, with an ancient castle, has extensive bleachfields, some of which were founded as far back as 1774. (p. 117.)

Inchture (545)*, Gaelic *Innis-tuir*, "island of the tower," in the Carse of Gowrie 14 miles east by north of Perth, occupies the summit of rising ground, which at one time must have been completely surrounded by water. Hence its name. Near Inchture is Rossie Priory. (pp. 122, 123, 140.)

Kenmore (1106)*, Gaelic *ceann-mhoire*, "Mary's headland," is a picturesque village at the eastern end of Loch Tay. Over the chimney piece of the inn parlour Burns wrote what has been pronounced by Lockhart as among the best of his English heroics. Wordsworth and his sister visited Kenmore in 1805. On an island in the loch opposite Kenmore there are the ruins of a priory, where Sibylla, daughter of Henry I of England, and consort of Alexander I of Scotland, was buried. (pp. 22, 23, 129, 134, 138, 143.)

Killin (1412)*, Gaelic *cill-Fhinn*, "Fingal's burial place," lies at the head of Loch Tay. Dr MacCulloch described the Killin neighbourhood as "the most extraordinary collection of extraordinary scenery in Scotland—unlike everything else in the country and perhaps on earth and a perfect picture gallery in itself, since you cannot move three yards without meeting a new landscape." At the upper end of the village a bridge of five unequal arches spans the Dochart. The view up the river from this point is very fine and has been painted by many artists. At Auchmore House, the seat of the Marquis of Breadalbane, may be seen the largest vine in the world. A monument has been erected in the village to the Rev. James Stewart, the first to translate the New Testament into Gaelic. There is a tweed factory in the village. (pp. 22, 41, 43, 82, 87, 106, 125, 138, 143.)

Lochearnhead, a small straggling village at the west end of Loch Earn, is much frequented by tourists. In the immediate vicinity are the Edinample falls and Glen Ogle. Behind the village is an interesting group of stones with cup and ring markings. (p. 106.)

Logierait (1618)*, Gaelic *lag-an-rath*, "hollow of the castle," lies on the north bank of the Tay about half a mile above its

Killin

junction with the Tummel. On an eminence near the village there has been erected a splendid Celtic cross to the memory of George, the sixth Duke of Atholl. Logierait was the seat of the Court of Regality in which the Dukes of Atholl administered feudal justice from the twelfth century to the abolition of hereditary jurisdiction. (pp. 143, 151.)

Longforgan (1997)*, in the Carse of Gowrie, commands a fine view of the whole Carse. The village consists of a straggling street, which formerly served as an appanage of Castle Huntly.

Luncarty, a village in the Strathmore district of Perthshire four miles north-west of Perth. Here the Danish invaders suffered defeat about the year 990. During the battle, according to the legend, Kenneth was greatly assisted by a peasant-ancestor of the Hays, who for his services obtained a large grant of land. (p. 93.)

Methven (1847)*, lying about six miles to the north-west of Perth, has in its neighbourhood several famous trees including the Pepperwell Oak, which, with a girth of over 15 feet, is known to be over 400 years old. One of the most celebrated places in the neighbourhood is Lynedoch Cottage, the scene of the touching story of Bessie Bell and Mary Gray. (pp. 94, 116, 138, 140, 150.)

Muthill (1269)*, three miles to the south of Crieff, was a seat of the Culdees in the twelfth century, and later the residence of the Deans of Dunblane. The old church, now a most interesting ruin, is said to have been erected by Bishop Ochiltree in 1430. (p. 153.)

Perth (36,669)*, both from an historical point of view and from the great beauty of its natural surroundings, is one of the most interesting towns in Scotland.

With the exception of the modern suburbs, it is almost entirely situated on a spacious plain lying but a few feet above the level of the river. It is bounded, north and south, by the fine meadows called the Inches, the name indicating that they were at one time islands. On the opposite side of the river Tay rises the Hill of Kinnoull, its sides highly cultivated and studded with elegant villas. On the west the slope is gradual and easy. The ascent on the south is more abrupt and forms the Hill of Moncrieff, Friarton and Craigie. Towards the north there is no elevated ground between Perth and the Grampian mountains from 10 to 12 miles away.

At a very early period Perth was walled and fortified, and

Perth, from Kinnoull Hill

girded by a ditch or fosse supplied with water from an aqueduct from the river Almond. The Castle was situated at the north-east corner of the town, and a high tower or turret stood at the West Port near what is now the junction of the High Street and Methven Street. A general idea of the extent and shape of ancient Perth will best be formed if we remember that the aqueduct still keeps its course to the Tay round what was formerly the base of the city walls. After the town was taken in

1651 by Cromwell, the fortifications were allowed to go to ruin. The only remaining part that can now be seen is that lying between George Street and Skinnergate.

Up till the year 1720 the town consisted simply of two long streets which ran parallel in an east and west direction—the High Street and the South Street.   Between these two streets and running off them were several narrower ones known as gates and

Tay Street, Perth

vennels.   The principal parts of the town were then situated in the neighbourhood of the Watergate and the Speygate.   The position now occupied by the Jail and County Buildings was the site of Gowrie House.   The Skinnergate, the Castle Gable and the Horse Cross were at that time the principal business centres. Many of the old houses stood a foot or two below the level of the street, and had arched doorways and windows.   On the front wall there was placed a superstructure of wood about six feet in breadth.

The ground-floor was open, forming "channels," as they were called, and it was here that the goods for sale were displayed. About 1760 the town began to be extended. (pp. 2, 23, 25, 49, 59, 64, 66, 68, 69, 78, 81, 82, 90, 92, 93, 100, 101, 102, 103, 113, 116, 125, 135, 137, 139, 141, 143, 145, 151, 153.)

Pitcairngreen, Gaelic *pitht-a-chairn*, "hollow of the cairn," on the left bank of the Almond four and a half miles north-west of Perth, is, like the other small villages in this neighbourhood, principally engaged in the bleachfields. When the village was founded towards the close of the eighteenth century, it was predicted that it would become a rival to Manchester.

Pitlochry, 350 feet above sea-level on the left bank of the river Tummel, 28½ miles from Perth, is much frequented on account of its salubrious climate and beautiful scenery. After leaving the village, the main road to the north winds through the Pass of Killiecrankie, one of the narrowest and most beautiful in Scotland. Though now possessing all the modern conveniences of life, yet at no distant date Pitlochry was a rude Highland village with only a few slated houses. (pp. 82, 101, 137, 138.)

Scone (2341)*, a flourishing village two miles north-east of Perth on the road to Blairgowrie, is known as New Scone to distinguish it from the hamlet of Old Scone, and dates from the beginning of the nineteenth century. A monument has been erected to the memory of David Douglass, the celebrated botanist, a native of Scone. The hamlet of Old Scone was situated about a mile to the west but it has now all disappeared with the exception of a fine old cross surrounded by lordly trees. In the eighth century Old Scone was the capital of Pictavia. There the Stone of Destiny, says tradition, was transferred from Dunstaffnage by Kenneth Mac Alpine. The Scottish princes were crowned on the Stone of Destiny until it was removed to Westminster in 1296 by Edward I of England, to form part of the English coronation chair. A legend was woven round the stone, which acquired a

Coronation Chair

sacred character as influencing the destinies of the Scottish nation. This was expressed in a Latin rhyme, which has been translated

> "Unless the fates are faithless grown
> And prophet's voice be vain,
> Where'er this monument is found
> The Scottish race shall reign.

The stone is identical in every respect with the sandstone rock of the neighbourhood, and the story is probably nothing more than a myth. Parliaments were often held at Scone. In the year 1841 part of the buildings of the Abbey of Scone was laid bare in the old burying ground. The Abbey was sacked and burned in 1559 by a mob from Perth. The "Moot Hill" is another interesting object situated within the Palace policies, from whence it is said the early Scottish kings promulgated their edicts. (pp. 85, 116, 128, 148.)

Stanley (1388), on the right bank of the Tay seven and a quarter miles north-west of Perth, grew up in connection with the cotton mills established by the Arkwrights. Stanley House was once the seat of the Lords Nairne and has many Jacobite associations. Within the Stanley policies is a remarkable round structure of great age, now in ruins. According to tradition it was a religious house in connection with the Abbey of Dunfermline, but it has more the appearance of a baronial fortalice. (pp. 81, 137, 138.)

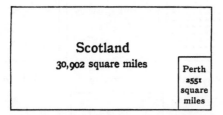

Fig. 1. Area of Perthshire compared with that of
Scotland

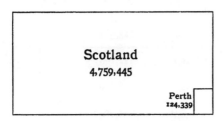

Fig. 2. The Population of Perthshire compared with
that of Scotland (1911)

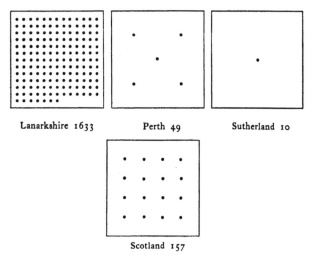

Fig. 3. Comparative density of Population to the square mile in 1911

(*Each dot represents 10 persons*)

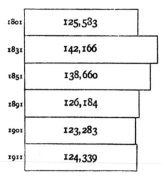

Fig. 4. Diagram showing increase and decrease of Population in Perthshire since 1801

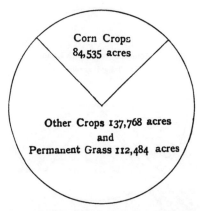

Fig. 5. Proportionate area under Corn Crops compared with that of other cultivated land in Perthshire (1910)

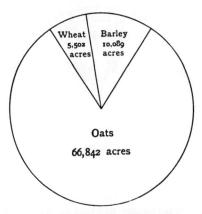

Fig. 6. Proportionate area of chief Cereals in Perthshire (1910)

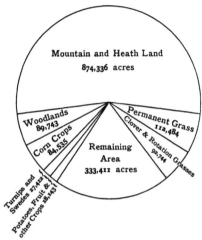

Fig. 7. Proportionate areas of land in Perthshire (1910)

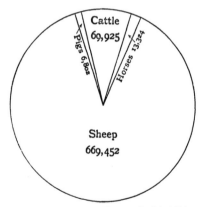

Fig. 8. Proportionate numbers of chief Live Stock in
Perthshire (1910)

www.ingramcontent.com/pod-product-compliance
Ingram Content Group UK Ltd.
Pitfield, Milton Keynes, MK11 3LW, UK
UKHW042143280225
455719UK00001B/69